VALUE MANAGEMENT IN CONSTRUCTION: A CLIENT'S GUIDE

John N. Connaughton BSc PhD ARICS
Davis Langdon Consultancy

Stuart D. Green BSc MSc CEng MICE MCIOB
The University of Reading

CONSTRUCTION INDUSTRY RESEARCH AND INFORMATION ASSOCIATION

© 1996 · ALL RIGHTS RESERVED

6 STOREY'S GATE · WESTMINSTER · LONDON SW1P 3AU

TELEPHONE 0171 222 8891 · FACSIMILE 0171 222 1708 · EMAIL SWITCHBOARD@CIRIA.ORG.UK

CONTENTS

THOSE WHO ARE UNFAMILIAR WITH VALUE MANAGEMENT IN CONSTRUCTION SHOULD READ PARTS 1 AND 2

ALL USERS SHOULD READ PART 3

BUILDING COST-EFFECTIVELY:
VALUE ENGINEERING

PART FIVE   33

VALUE MANAGEMENT IN A CLIENT ORGANISATION

PART SIX   45

THE VALUE MANAGEMENT TOOLBOX

APPENDICES   51

WHY READ THIS GUIDE ?

PART ONE

1.1 WHAT IS VALUE MANAGEMENT?

Value Management helps clients ensure that their investment in construction produces valuable assets which are cost effective to construct, use and maintain. It is a structured approach to defining what 'value' means to a client when meeting a perceived need, and to delivering that value via the design and construction process. It does this by clearly defining and agreeing project objectives and the means of achieving them.

1.2 WHY USE VALUE MANAGEMENT?

The decision to build is a major one, with far-reaching consequences. It involves a substantial investment of time, effort and money, and produces valuable assets which affect the lives and businesses of those who own and use them. Value Management is attracting considerable interest in UK construction as a means of improving the overall value and performance of construction projects. This guide shows how Value Management may be used to do this and, in particular, to:

- identify and evaluate the need for construction before making a major financial commitment
- identify and prioritise key project objectives
- ensure all aspects of the project design are the most effective for their purpose in terms of cost, quality and satisfaction of the client's long-term needs
- improve team building
- eliminate unnecessary cost.

1.3 WHO SHOULD READ THIS GUIDE?

[1] **Potter, M. (1995)** *Planning to Build?*, CIRIA. Other companion guides available from CIRIA are listed in *Toolbox 6.*

This guide is for all who build, be they clients, their advisers or contractors. Users are expected to have some knowledge of the construction process, the participants typically involved and of the different procurement arrangements available. First-time clients are referred to the CIRIA publication *Planning to Build?*[1].

1.4 COVERAGE

There is no single prescription for using Value Management. This guide sets out the essential principles of good practice as applied to construction, including both building and civil engineering. It outlines the main approaches for a range of clients and for different types of projects.

While it is for the user to apply these principles in particular circumstances, the guide will help to:
- identify when to use Value Management
- understand the implications of using Value Management and the results which can be expected
- select the most appropriate approach
- put in place procedures to get the most from Value Management.

1.5 HOW TO USE THIS GUIDE

This guide is in six main parts (including this introduction – **PART 1**):

PART 2 introduces the main uses and benefits of Value Management and what can be achieved. *Those already convinced of the merits may go straight to Part 3.*

PART 3 provides an overview of Value Management, outlining its core principles, when it may be used and what is involved. This part provides a framework for the systematic application of Value Management for different clients and types of construction projects. *All users should read Part 3.*

PART 4 outlines how Value Management can deal with projects which are particularly difficult to define because of the nature of the project itself and the range of 'clients' it must satisfy.

PART 5 outlines how Value Engineering can deal with projects which are easier to define and where there is a good deal of agreement over their main objectives and priorities.

PART 6 outlines how Value Management can be implemented within a client organisation. This is of particular relevance to clients who build regularly and who wish to ensure that they achieve value for money on all their projects.

THE VALUE MANAGEMENT TOOLBOX provides appendices which describe the essential techniques for those applying them. It also provides useful checklists and identifies sources of further information.

1.6 BASIS OF GUIDANCE

The advice contained in this guide is based on a study undertaken for CIRIA between January and September 1995. Its objectives, approach and findings are presented in the final report *Value Management in UK Practice*, available from CIRIA (see *Toolbox 6*).

1.7 ACKNOWLEDGEMENTS

This publication was funded by:

> Dearle & Henderson
> Department of the Environment (Construction Sponsorship Directorate)
> Highways Agency
> Southern Water Services Ltd
> South West Water Services Ltd
> Thames Water Utilities Ltd
> CIRIA Core Programme.

The research and preparation of this guide was commissioned by CIRIA and undertaken by Davis Langdon Consultancy in association with the University of Reading. The research team comprised:

J N Connaughton	Davis Langdon Consultancy
S D Green	The University of Reading
C Truniger-Johnston	Davis Langdon Consultancy
with:	
B Nivison	bruce[n] (Design and Layout)
J Campion	Quintessential Communications (Editorial)
M Potter	Davis Langdon Consultancy (Cover drawing)

The work was overseen by a Steering Group comprising:

S H Walker	Taylor Woodrow Management Ltd (Chairman)
L Edwards	GMW Partnership
J Jayasundara	Highways Agency
P J Mason	Sir Alexander Gibb & Partners Ltd
S Mitchell	Ove Arup Partnership
P A Popper	Higgs & Hill Construction Holdings Ltd
R Poynter-Brown	Dearle & Henderson
M Quarterman	Value Management Ltd
K Vanstone	South West Water Services Ltd
A J Walters	Mott MacDonald Foundations & Geotechnics
P J Whatley	Southern Water Services Ltd
J Whiteman	Value Management Ltd

Valuable assistance was also provided by:

G H Bateman	Thames Water Utilities
A Boyle	Highways Agency
C T Cain	Defence Estate Organisation (Works)
J Gravett	Gardiner & Theobald
V J A Hughes	Rofe, Kennard & Lapworth
D Sugg	Fuller Peiser Property Consultants
P V Lawrence	Frank Graham Consulting Engineers Ltd
C Parsloe	BSRIA
R L W Tyler	Acer Consultants Ltd
D Woodhouse	BAA Plc

The project was managed on behalf of CIRIA by:

A Jackson-Robbins	(until April 1995)
and by:	
R Bishop	CIRIA's Research Manager for Management Topics, from April 1995.

A large number of people and organisations representing a broad spectrum of views participated in the study. A total of 12 people were interviewed formally by the consultants and another 77 individuals and firms responded to a postal survey. A list of participants is provided in the study final report *Value Management in UK Practice*, available from CIRIA (see *Toolbox 6*). Their contribution is gratefully acknowledged, as is that of the representatives of the major professional bodies who participated in 'peer review' by reading and commenting on early drafts of this guide.

WHY USE VALUE MANAGEMENT ?

PART TWO

2.1 WHAT CAN BE ACHIEVED?

**THE MAIN
BENEFITS**

2.1.1 This guide will help clients understand what Value Management involves and what resources and other support are needed to exploit fully its potential benefits.

Properly organised and executed, Value Management will help clients achieve *value for money* from their construction projects by ensuring that:
- the need for projects is always verified and supported by data
- project objectives are openly discussed and clearly identified
- key decisions are rational, explicit and accountable
- the design evolves within an agreed framework of project objectives
- alternative options are always considered
- outline design proposals are carefully evaluated and selected on the basis of defined performance criteria.

Value Management can also provide other important benefits:
- improved communication and teamworking
- a shared understanding among key participants
- better quality project definition
- increased innovation
- the elimination of unnecessary cost.

**REDUCING COSTS
AND ACHIEVING
VALUE FOR MONEY**

2.1.2 By eliminating unnecessary cost, capital cost savings of between 10% and 25% can typically be made on construction projects. While these savings more than offset the additional costs of Value Management – estimated at between 0.5% and 1% of project costs – they should not be seen as an end in themselves. The emphasis in this guide is on achieving maximum value for the resources available. This will provide far greater benefits over the lifetime of a project than even substantial savings in capital cost.

The key differences between Value Management and cost-reduction are:
- Value Management is positive, focused on value rather than cost. It seeks to achieve the best balance between time, cost and quality.
- Value Management is structured and accountable.
- Value Management is multi-disciplinary and seeks to maximise the creative potential of all project participants working together.

In any event, Value Management is just as valuable when it confirms that, for instance, the original design proposals are the most effective as when it discovers better alternatives.

66 *The business has looked hard at the way in which its projects are initiated and implemented and found considerable scope for the application of Value Management throughout the project delivery process. Against this background its Directors and Senior Managers have identified, in particular, that Value Management is a means by which resources can be optimised in an endeavour to achieve the business's target of a 20% reduction in project delivery costs over 10 years without sacrificing quality or safety considerations.* **99**

Source: Presentation by Ian Fry, Senior Project Manager, LUL, to a conference organised by 'IBC Legal Studies and Services', 28th June 1994.

VALUE
MANAGEMENT
AND VALUE
ENGINEERING

2.2 VALUE MANAGEMENT DEFINED

2.2.1 Everyone involved in construction should be concerned with value for money. Any construction project has limited resources, and success (or failure) depends on how well these resources are used. Value Management is a strategy for identifying the project that gives best value for money and which makes full use of these scarce resources. It is essentially project-wide in scope and is concerned with identifying and meeting project needs within time and cost constraints.

> ❝ *Value Management is an approach which aims to establish, at the start of the project, the strategic plan by which it should develop. This is partly achieved through the use of a series of workshops at key stages of the project; and is complemented by Value Engineering techniques. These also make use of workshop techniques, but are concerned with obtaining value-for-money through an organised systematic approach, placing emphasis upon whole-life costing.* ❞

Source: **Highways Agency (1996)** *Trunk Roads and Motorways:*
Review of Contractual Arrangements, Consultation Document, January.

A range of terms is currently used within UK construction to describe different parts of this strategy. In this guide, two terms are used as follows:

VALUE MANAGEMENT which is defined as:
A structured approach to defining what value means to a client in meeting a perceived need by establishing a clear consensus about the project objectives and how they can be achieved.

It also incorporates **VALUE ENGINEERING,** which is defined as:
A systematic approach to delivering the required functions at lowest cost without detriment to quality, performance and reliability.

Value Engineering is therefore a special case of Value Management (Figure 1). The distinction between these terms is explained more fully in Part 3.

FIGURE 1   RELATIONSHIP OF VALUE MANAGEMENT AND VALUE ENGINEERING

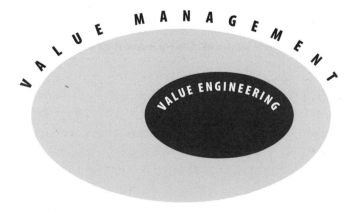

2.3 VALUE FOR MONEY IN CONSTRUCTION

WHAT IS VALUE FOR MONEY?

2.3.1 There is a commonly held misconception that value for money in construction can be achieved simply by building cost-effectively. But an effectively built project may fail to provide the necessary level of performance. Value for money takes account of the requirements of those whom the project is for, as well as the money available to construct it.

In practice, it can be difficult to define clearly and precisely how a project will meet the requirements of those who are to benefit from it. Part of the problem is that projects are frequently needed by a range of 'stakeholders'[2], each of whom may well have very different ideas of what (and who) it is for. Also, the financial resources available for construction projects are usually limited. So even if all possible project objectives could be clearly defined and agreed, the problem remains of how to build in the most cost-effective way.

Value for money is the optimum combination of whole life cost and quality to meet the client's requirement and depends on:
- agreeing an unambiguous set of project objectives, *and*
- ensuring that they are achieved cost-effectively.

Value Management helps ensure value for money by addressing both of these in turn.

[2] The term 'stakeholder' is used throughout this guide to refer to investors, end-users and others with a real interest in - and the power to influence - the project outcome. A glossary of key terms is given in *Toolbox 1*.

ESTABLISHING OBJECTIVES AND BUILDING EFFECTIVELY

2.3.2 Any project should be initiated only after a careful analysis of need. Failure to do this will cause problems with design and construction. More importantly, it will also cause problems with operation and use in the longer term.

Stakeholders' needs, expectations and aspirations are not always mutually compatible, nor are they consistent with limitations on time and money. A careful analysis of need is much more than a simple 'wish list' of all possible requirements. It involves agreement among key stakeholders about project objectives, giving due weight to their relative value and importance.

Value Management works by making explicit stakeholders' objectives and value for money criteria. This helps to investigate fully the need for a project before making a financial commitment. It provides a structured framework within which subsequent decisions can be taken in accordance with these objectives and criteria. This means that the project design is developed and continuously evaluated against need so that value for money is achieved.

Value for money can be improved either by enhancing the requirements, or by reducing the cost of meeting them. Indeed, it is often worth spending more money for a higher level of stakeholder satisfaction. The search for value for money is therefore trying to find the best balance between meeting the requirements of the stakeholders and the resources available. ***It is not simply about reducing costs.***

2.4 ACHIEVING VALUE FOR MONEY

IS VALUE MANAGEMENT NOT WHAT PROFESSIONALS DO ANYWAY ?

2.4.1 Some construction professionals argue that Value Management 'is what they do anyway'. In some cases this may well be true; in others less so. While existing practice may well achieve value for money some of the time, the aim must be to achieve it *all the time*. To do this clients must adopt a planned and rigorous approach to providing value for money. Value Management is just such an approach.

THE IMPORTANCE OF A PLANNED APPROACH

2.4.2 By using Value Management, clients can feel confident that they will achieve value for money. There are a number of key points in every construction project at which the client must make important decisions. Value Management helps to ensure that these decisions are taken in a way which is:

- rational
- explicit
- accountable and auditable.

When the project is complete clients will have an auditable record of key decisions to show how value for money was delivered. But construction is an inherently uncertain process and Value Management cannot guarantee a successful outcome. However, decisions made in this way will need less revision and therefore eliminate a major cause of disruptive change. Needless variation orders and other disruptions militate against achieving value for money.

DECISIONS NEED TO BE RATIONAL

2.4.3 Value for money in construction depends on rational decisions. This means ensuring that:

- the nature of the problem is fully understood
- decisions are made in the light of agreed objectives
- different options for achieving the agreed objectives are considered
- the options (and their associated risks) are carefully thought out
- decisions are made on the basis of the best available data
- decisions draw on the widest possible range of expertise.

DECISIONS NEED TO BE EXPLICIT

2.4.4 It is not, however, sufficient for key decisions to be rational. They must also be made in a way that is clear and understandable. A structured approach to decision making – such as that provided by Value Management – makes the process explicit. This means that key project stakeholders are able to participate more fully and usefully in decision making, which increases their confidence in the process. It also means that they are more committed to a successful project outcome.

In other words, Value Management helps the stakeholders 'buy-in' to the project. By involving them in the design process, from which they are frequently excluded, they obtain a project that better meets their needs and with which they are satisfied.

DECISIONS
NEED TO BE
ACCOUNTABLE

2.4.5 Project decisions not only need to be rational, they also need to be seen to be rational. Clients are increasingly concerned about being able to justify their decisions after the event. Properly planned and documented, Value Management can provide clear evidence of rational and explicit decision-making against value for money criteria. This is important as it provides an audit trail and also helps manage continuous improvement in the context of Quality Management.

KEY SUCCESS
FACTORS

2.4.6 For Value Management to be effective, clients must:
- be committed at a senior level to its introduction and implementation
- ensure that it addresses their strategic business objectives
- be well informed about when to use it, what to expect and who to involve, including how to:
  - ensure that the key stakeholders are involved
  - identify and appoint experienced and effective personnel
  - ensure that design team members are fully briefed about their role in Value Management.

Subsequent sections of this guide show how to plan for Value Management and describe what clients need to do to ensure that value for money is being obtained at each key stage.

GETTING STARTED: AN OVERVIEW OF VALUE MANAGEMENT

PART THREE

3.1   THE CORE PRINCIPLES

A CREATIVE APPROACH TO PROBLEM SOLVING

3.1.1   Value Management uses a set of creative problem-solving techniques, in a team environment, to evaluate rigorously key project decisions. The basic stages of creative problem solving are widely recognised as follows:

- define the problem
- identify different options for resolving it
- evaluate the options
- select the option offering best value for money.

The Value Management approaches described in this guide are all based on these stages. While good design practice also follows them, the advantage of Value Management is that each stage is identified and performed *separately.* This ensures that the creative process of identifying possible options is kept separate from evaluation. This is important because a wide range of possible solutions must be considered to ensure value for money – the first solution to appear workable is not always the best.

THE IMPORTANCE OF WORKSHOPS

3.1.2   Creativity and imagination are needed to find the best value for money solution for stakeholders' needs. In practice, a particularly good way of unlocking this creativity is to hold a *workshop* (or more usually a series of workshops) involving a team of key stakeholders and other project participants.

Properly managed workshops - where people work together to find the best value for money solution to a particular problem or situation - are so effective that they are the central element in most approaches to Value Management.

Key features of such workshops are that they:

- coincide with key decision points in the project
- intervene in the design/construction process.

THE PLANNING AND TIMING OF WORKSHOPS AND THE NEED FOR ADVICE

3.1.3   The importance of a planned approach to Value Management has already been noted (Part 2). On construction projects, this is typically provided by planning a series of workshops to help define, develop, challenge and subsequently evaluate the project design proposals.

The *timing of workshops* is very important. They must be tied closely to key stages in project development if they are to provide the best opportunity for identifying needs and challenging key design decisions before they are made. As key decisions affecting project value are taken early in the project, Value Management is most useful in the early stages of project development (see Figure 2 opposite).

FIGURE 2   OPPORTUNITY TO IMPROVE VALUE FOR MONEY

CONCEPT	FEASIBILITY	SCHEME DESIGN	DETAIL DESIGN	CONSTRUCTION ➤

COST ➤

OPPORTUNITY
TO
IMPROVE
VALUE

INFLUENCE ON TOTAL VALUE

COST OF LATE CHANGES

VALUE ➤

T I M E ➤

The ***number of workshops*** is also important. Too many, and the design and construction process may be disrupted and delayed. Too few, and opportunities for improving the value of design proposals may be lost.

In any construction project there are many opportunities to influence and review the development of the design. Experience shows that, typically, there are a small number of points where key client decisions must be made and where workshops are particularly effective. The more obvious are shown on a simple project scheme in Figure 3. These points may be linked to the client's capital expenditure approval procedure and are often programmed in advance as key design management milestones.

FIGURE 3   OPPORTUNITY POINTS FOR VALUE MANAGEMENT WORKSHOPS

CONCEPT   OPPORTUNITY   FEASIBILITY   OPPORTUNITY   SCHEME DESIGN   OPPORTUNITY   DETAIL DESIGN   OPPORTUNITY   CONSTRUCTION

The key opportunity points are typically:
- during the concept stage, to help identify the need for a project, its key objectives and constraints
- during the feasibility stage, to evaluate the broad project approach/outline design
- during scheme design, to evaluate developing design proposals
- during detailed design, to evaluate detailed design proposals.

Not all projects follow this pattern. The scale and complexity of the project will determine the number of workshops, and when they should be held. But the client should identify the key decision points so that workshops can be planned in advance. Advice from a Value Management expert will be needed to do this, and to organise and manage the workshops.

❝ *Once you've decided what the customer wants, then you have to have some idea as to what is the best practice answer to that kind of mixture of customer requirements.*

At this point we carry out an exercise called 'Value Management One', which asks: 'Is this worth doing, have we got the cost right to plus or minus 20% and is the project still a good one if it is going to cost 20% more than the mainstream cost?'

If the project passes VM1, it moves to the next stage, the choice of the 'right alternative'. A project team is created, including an architect, a construction manager and many of the large sub-contractors. The team starts by assessing the 'alternatives' for realising the building. This is followed by 'Value Management Two', in which the chosen alternative is examined. At this stage costs are narrowed down to +/- 10%. ❞

Source: Sir John Egan, Chief Executive, BAA plc
New Builder, 8th April 1994.

The Value Management expert may be a specialist **value manager,** or may also be a construction professional such as a design or cost consultant, construction manager, project manager or contractor. The value manager may also be asked to chair or 'facilitate' workshops, and in this case is known as a **facilitator.** Whoever they employ as value managers or facilitators, before appointing them clients should make sure that the individuals proposed have the required skills, expertise and experience. The key criteria in the selection of value managers and facilitators are summarised in Toolbox 2.

THE DESIGN OF WORKSHOPS

3.1.4 The **design of workshops** is determined primarily by their timing and objectives. There is little point, for instance, in holding a workshop to help develop design proposals at a very early stage when insufficient information is available. The key to a successful workshop lies in giving it a clear objective. The format, participants and techniques to be used all depend on whether the aim is to identify and agree project objectives, for example, or to find the most cost-effective design. **Indeed, Value Management depends fundamentally on whether or not stakeholders can agree project objectives from the start. This distinction is crucial and it is important that users of this guide adopt the appropriate approach.** Part 3.2 provides further details.

PARTICIPATION IN WORKSHOPS

3.1.5 Workshops should be organised and managed by an experienced facilitator. The involvement of key project stakeholders, particularly in the workshops at the concept and feasibility stages is critical and needs to be planned for. Parts 3 and 4 review the roles of the various project participants in different Value Management approaches.

In some cases a Value Management team, separate from the project design team, may be appointed to give an independent review of all aspects of the design at the scheme design stage. This should not be necessary, however, if earlier workshops have been effective. It is also relatively costly and may lead to conflict with the project design team. While there are circumstances in which it may be appropriate to use an external team, it is not common practice in the UK. The benefits of using the existing design team - in terms of, for example, project familiarity and team-building - are generally felt to outweigh the value of an independent critique of the design. This is dealt with in more detail in Parts 4 and 5.

VALUE MANAGEMENT AS A CONTINUOUS PROCESS

3.1.6 Although Value Management is centred on a series of workshops, it is in fact an ongoing process. Workshops must not become a 'safety net' to finalise incomplete elements of the design or to rectify mistakes. Instead, the structure provided by planned workshops should be used to reinforce the need to develop and review the design against client needs continuously throughout the project. Clients and their project managers should encourage a Value Management 'culture' or philosophy which seeks to maximise value for money and eliminate waste at every opportunity. Parts 4 and 5 provide further details.

3.2   HOW TO USE VALUE MANAGEMENT: A PLAN OF ACTION

FINDING THE RIGHT APPROACH

3.2.1 Different approaches to Value Management suit different situations. For example, there would be little point in using workshops to focus on building cost-effectively if there was no agreement about project objectives. Before value for money can be examined - and achieved - an unambiguous statement of project objectives is needed. In some cases this will already exist. In others, it can be difficult to define and agree. This is especially true of projects which must satisfy a large number of different interest groups.

It is important, therefore, to distinguish between projects where objectives are clear and shared by the key project stakeholders and those where objectives are ambiguous and not so well shared. While this distinction is not clear cut in practice, it is useful in finding the right approach, as the aim of Value Management depends fundamentally on whether objectives are clear or ambiguous.

Reaching an agreement for projects that have clear objectives which are shared by their key stakeholders is relatively easy. In this case the ***aim of Value Management is to achieve the project objectives in the most cost-effective way.*** Examples of these projects are shown in Boxes 1 and 2 on page 16.

For projects with ***ambiguous objectives*** which are not shared by the key project stakeholders, there may be different interpretations, not only of the project objectives but of the need which the project is trying to satisfy. In this case, value for money depends on achieving an unambiguous and agreed set of objectives and ***the aim of Value Management is to resolve differences about project objectives and achieve a common consensus among stakeholders.*** Examples are provided in Boxes 3 and 4 on page 17.

FIGURE 4   VALUE MANAGEMENT FOR DIFFERENT PROJECT CIRCUMSTANCES:
CLEAR OBJECTIVES, SHARED BY STAKEHOLDERS

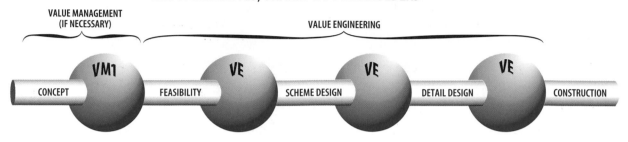

BOX 1

Clear objectives: The Speculative Development

Speculative developers usually have clear and precisely defined project objectives, e.g. maximising lettable floor space for minimum construction cost and time. They do not normally have to deal with different internal interest groups each pursuing their own objectives.

Such clients will find Value Engineering particularly useful. As there is no need to resolve ambiguity about project objectives, they can use Value Engineering from the early project stages. The emphasis is on enhancing the potential revenue from the project, reducing the project cost or timescale, or both. The strategy described in Part 5 of this guide would be of benefit to these clients during the design and construction stages.

BOX 2

Clear objectives: The Civils Project

Many public projects – including civil engineering projects – have ambiguous objectives which may not be well shared among key stakeholders (see Box 4 opposite). However, civils projects often have very clear and well-defined objectives. For example, there may be little disagreement among stakeholders about the objectives (defined in technical performance terms) of a retaining wall, a railway bridge or a section of motorway. In this case, the civils client can, like speculative developers, use Value Engineering from the early project stages, in line with the strategy described in Part 5 of this guide.

Value Engineering seeks to improve the cost-effectiveness of the project design, when project objectives are confirmed. Design solutions are evaluated in terms of risk, cost, technical performance and life-cycle issues. Particular attention may be paid to 'buildability' on these projects, as a relatively high proportion of their total cost relates to temporary works and the construction process.

However, Value Engineering can only be used on these projects if the relevant policy issues, including project location, have been resolved. A variety of different groups may, for example, have an interest in decisions about the routing of major highways. While Value Management as described in Part 4 can be useful in resolving potential conflict between these groups, there will not always be sufficient common ground for them to reach an agreement. Progress may then depend on executive decisions which leave some interest groups dissatisfied.

FIGURE 5   VALUE MANAGEMENT FOR DIFFERENT PROJECT CIRCUMSTANCES:
AMBIGUOUS OBJECTIVES, NOT SHARED BY STAKEHOLDERS

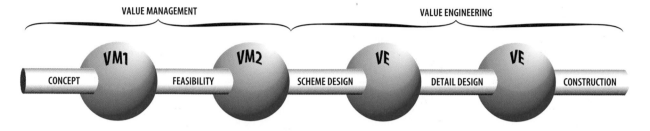

BOX 3

Ambiguous objectives: Multi-User Research Laboratory

A good example of a project with potentially ambiguous objectives which are not well shared by key stakeholders is a multi-purpose laboratory building. Senior managers may place different values on the use of the laboratory as a show case for visiting clients and as a functioning laboratory. Research scientists may want the building to keep particular research groups separate; others may argue that researchers need to work closely together to encourage the exchange of ideas. Those responsible for operation and maintenance will have their own design requirements and priorities.

In this case, the project objectives cannot be determined in advance. Value for money will depend crucially on the key stakeholders developing a shared understanding of the project objectives. This may be difficult to achieve and involve a 'trade-off' between the requirements of different groups, separating 'needs' from 'wants'. This is only possible if stakeholders 'buy in' to the process. The approach to Value Management in Part 4 of this guide is particularly applicable in such circumstances.

BOX 4

Ambiguous objectives: The Public Sector Building Project

Many public sector projects cater for a diverse range of interest groups. For projects with ambiguous objectives, value for money is not simply about building cost-effectively. It depends crucially on the key stakeholders first of all agreeing an unambiguous set of project objectives.

A common difficulty with public sector projects is knowing who the stakeholders are. To say that 'everybody in the community is a stakeholder' does not lead to effective decision-making. Stakeholders must be found who represent the various interest groups and who will stand by their decisions after the event. Moreover, decision making must be accountable. Not only must value for money be achieved, it must be seen to be achieved.

The Value Management strategy described in Part 4 will help identify the key project stakeholders and provide a basis for decision making which is open, rational and accountable.

This can only be done with the active participation of the key stakeholders. Stakeholders include the parties who will evaluate the success of a project on completion. Only if they are involved in the Value Management process, can project objectives be achieved effectively.

This distinction helps decide which Value Management approach is best suited to the client and project circumstances, ***particularly in the early project stages.***

A FRAMEWORK FOR VALUE MANAGEMENT

3.2.2 Where objectives are ambiguous and not well shared, differences must be resolved if the project is to be realised. This first stage of Value Management must be successfully completed before any attention can be given to building cost-effectively. Where objectives are clear, this stage may not be necessary. Instead, it may be more relevant to question and examine the need for a project in the first place.

Figures 4, 5 and 6 illustrate these contrasting situations and are essential in understanding the structure of the guidance presented here. The terminology currently used to describe these situations is subject to wide interpretation. ***This guide distinguishes a special case of Value Management which aims to achieve project objectives cost-effectively. This is called*** VALUE ENGINEERING. ***Other aspects of the strategy, including the first stages aimed at resolving multiple project objectives and occurring in the early stages of design, are called*** VALUE MANAGEMENT.

USING THE FRAMEWORK

3.2.3 To determine whether a particular project has clear or ambiguous objectives for the purposes of using Value Management, the client should try to answer the following questions:
- is there a consensus among stakeholders about what should be done?
- are the objectives predetermined?
- are the objectives likely to remain constant over time?

Only if the answer to each of the above questions is 'yes' should Value Engineering be used. Clearly, it is only possible to determine how to build cost-effectively if there is agreement about what should be done. When objectives are ambiguous the different views must be resolved before using Value Engineering. Equally, there is little point in using Value Management - which seeks agreement through discussion and debate - if there is no underlying willingness amongst the stakeholders to agree.

Figure 6 can be used to help determine an appropriate Value Management strategy.

FIGURE 6   A STRATEGY FOR ENHANCING VALUE

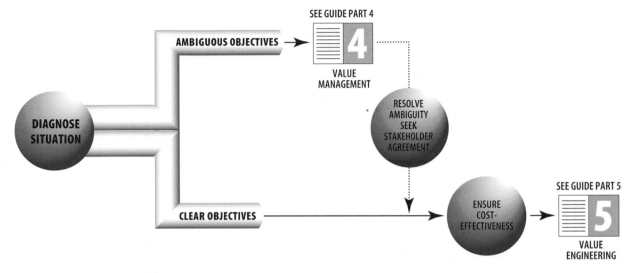

It is important not to confuse the broad strategy of Value Management with the narrower approach of Value Engineering. Value Engineering is a special case of Value Management which seeks to achieve a stated objective in the most effective manner. In essence, it is a technical activity involving the design and construction teams. By contrast, Value Management seeks to define the project objectives with the key project stakeholders in the early project stages.

3.3 PROJECT UNCERTAINTY AND RISK

WHAT IS RISK ?

3.3.1 Risk is the chance of an adverse event occurring. It can arise in construction from a variety of sources such as unforeseen changes in political and economic conditions, regulations affecting construction, human error, etc. It can affect the construction process, the health and safety of those involved and the wider environment. Although risk can never be removed entirely, it should not be left to chance. A systematic approach to risk management will:

- make risks explicit and easier to manage
- identify and assess the major risks
- put in place a means of dealing with these risks so that they are controlled and the potential damage is minimised.

[3] **Godfrey, P. (1995)** *Control of Risk,* CIRIA. See *Toolbox 6.*

A detailed treatment of risk is beyond the scope of this guide. A companion CIRIA publication provides detailed guidance to good practice in the management of risk on construction projects[3].

RISK
MANAGEMENT
AND VALUE
MANAGEMENT

3.3.2 Risk Management, Value Management and Value Engineering are all part of a single management structure. It is important, however, to differentiate between them so that the right techniques are introduced at the right time. Risk management is mainly concerned with events which might affect the ***achievement*** of project objectives. It requires objectives to be well defined – one cannot assess whether project objectives will be adversely affected unless there is a prior statement of what they are. Risk Management (and, in particular, risk identification and analysis) therefore has a vital role to play in identifying and choosing between competing technical solutions, which is the subject of Value Engineering (see Part 5).

Risk Management is also an important part of Value Management, even though it may seem unhelpful to try to identify and manage risks until there is agreement about what the objectives are. In fact, a strategic diagnosis of the risks may well influence how the objectives are set. A consideration of project risks is likely to feature in outline design proposals during project feasibility (Value Management at Feasibility, Part 4.4, page 28).

WORKSHOPS

4 For further details
see **Godfrey, P. (1995)**
Control of Risk, CIRIA.
See *Toolbox 6.*

3.3.3 Decisions about project objectives and preferred outline designs are strongly influenced by the risks of available alternatives. As workshops[4] are a useful way of identifying project risk, Value Management and Risk Management can be combined in the same workshop. Indeed, it is difficult truly to separate them. But workshops are demanding and it is important not to overload the participants by trying to achieve too much (and risking achieving nothing at all). If separate workshops are to be held, then Value Management workshops should normally precede Risk Management workshops.

AGREEING CLEAR OBJECTIVES : VALUE MANAGEMENT

4

PART FOUR

> 66 *Time and effort will be dedicated at an early stage to establishing the objectives, requirements and costings of the project. Value Management techniques will be used to test and justify them. Failure to undertake this process thoroughly could lead to problems later. Inadequate project definition is a major cause of failure in the public sector.* 99
>
> *Source:* **Chancellor of the Exchequer (1995)** *Setting New Standards - A Strategy for Government Procurement,* HMSO.

4.1 INTRODUCTION

This part of the guide deals with using Value Management to identify and agree project objectives. It is particularly relevant for projects where objectives are ambiguous and not well shared by key stakeholders. It also outlines how the best value for money approach to the project can be identified and selected.

It is important to recognise that Value Management draws on a range of methods and techniques. The following guidance, though linked to important project stages, is not a standard procedure to be followed precisely. The value of the approach lies in challenging statements of project objectives and intentions. If it does not stimulate thought, debate and innovation then it has failed.

4.2 PREPARATION

KEY ACTIVITIES   **4.2.1** The client's responsibility for developing a plan for Value Management has previously been noted (see 3.1.3). The plan is likely to be structured around workshops.

The key client activities are to:
- select and brief value manager (and/or workshop facilitator)
- provide an initial statement of need
- identify stakeholders
- organise and participate in workshop(s)
- report, implementation and feedback.

Each of these activities is dealt with below.

SELECT AND BRIEF   **4.2.2** The client needs a value manager to help develop and implement a Value Management
VALUE MANAGER   plan. The value manager is responsible for organising all aspects of the Value Management study and may also facilitate the workshop(s). He or she may come from within the client organisation or from an external firm. When appointing an in-house value manager it is important to select someone who is not directly involved in the project. This helps to ensure neutrality and a fresh, enquiring approach that challenges existing ideas and solutions.

The value manager should be clearly briefed regarding responsibilities for:
- collecting all information relevant to the Value Management studies
- selecting a Value Management team
- organising and facilitating the Value Management studies/workshops
- producing workshop reports for the client.

Value managers should have a good understanding of the construction process. Most importantly they need to have thorough knowledge and experience of Value Management and associated techniques. If they are to facilitate workshops, they also need to be competent in a range of management and interpersonal skills, including:

- the organisation and facilitation of brainstorming sessions/workshops
- the ability to seek innovative solutions
- the ability to motivate project participants to achieve project objectives[5].

5 *Toolbox 2* provides further details.

INITIAL STATEMENT OF NEED

4.2.3 Before the first Value Management workshop, the client should draw up a statement of why the project exists. This will be used to focus the 'information' stage of the workshop aimed at defining the project objectives. Workshop steps are explained in 4.3.2.

IDENTIFY STAKEHOLDERS

4.2.4 An important preliminary activity is identifying the project stakeholders. This can be difficult, but it is vital that all the key interest groups with power to influence the project outcome are represented. Stakeholders must be represented by senior individuals who can make decisions on their behalf and who will participate actively in the workshop. It is often useful to ask 'Who will judge the success of the project after completion?' when identifying key project stakeholders.

In many cases the number of stakeholder representatives will have to be balanced against the overall size of the Value Management team. While workshops with more than 12 participants can be difficult to manage, up to 25 or 30 stakeholders may be needed on large and/or complex projects. In this case a series of linked workshops may be better to ensure that all stakeholders are adequately involved and represented.

VALUE MANAGEMENT AWARENESS

4.2.5 It will usually be beneficial to run Value Management 'awareness' sessions/seminars, particularly where project participants are unfamiliar with the approach. These seminars perform a number of important functions. They serve as 'induction courses' for individuals who are unfamiliar with Value Management. They are also a useful way of raising an awareness of Value Management among key individuals within the client organisation and gaining their commitment to the approach, thereby creating a Value Management 'culture' within the firm (see also Part 6).

Awareness sessions should be chaired by a facilitator and attended by those who will participate in the Value Management studies. These sessions may last up to half a day and should cover:

- Value Management objectives and techniques
- how Value Management will be applied on particular projects
- information requirements for Value Management workshops and studies
- arrangements for management and communication.

WORKSHOPS

4.2.6 It is generally recommended that two workshops are held as part of a Value Management study: one during the concept stage (VM1) and one during the feasibility stage (VM2) – see 4.3 and 4.4. However, the precise timing and number of workshops depends on the scale and complexity of the project. The value manager provides specific guidance on this aspect of the Value Management Plan.

An agenda and briefing document (covering the purpose of the workshop, tasks and timetable) should be drawn up and circulated to all participants before the workshops. The VM1 briefing document should include the initial statement of need by the client and any information considered to be relevant, such as provisional cost information, initial drawings and details of risks and constraints.

6 *Toolbox* 3 provides further details.

The VM2 briefing document should also include the feasibility study and investment appraisal, a statement of objectives and constraints, the outline programme, cost report and drawings and details of any procurement proposals[6].

REPORT, IMPLEMENTATION AND FEEDBACK

4.2.7 After each Value Management workshop, the facilitator presents a report to the client. If the recommendations arising from workshops are to be implemented successfully, it is important to ensure that:
- a timescale for implementation is agreed
- a forum for discussing recommendations is clearly identified (in most cases, the existing regular project team meetings provide a suitable forum)
- individuals responsible for implementation are adequately briefed.

Finally, if the workshop is to form part of a wider, long-term Value Management strategy, it is important that feedback is obtained on the success or failure of the approach used. These issues are covered in Part 6.

4.3 VALUE MANAGEMENT AT CONCEPT STAGE (VM1)

OBJECTIVES

4.3.1 The primary purpose of Value Management at concept stage is to ensure that the ***need*** for a new project is thoroughly analysed before the client is committed to build. This will usually take the form of a workshop (VM1). The specific objectives of VM1 are:
- to agree clear project objectives and ensure that they are understood by all parties
- to state clearly the value for money criteria
- to provide useful ideas about possible options
- to ensure that the decision-making process is accountable.

KEY ACTIVITIES   **4.3.2**   The VM1 workshop is structured around the key stages of creative problem solving (see 3.1.1, page 12). A plan for this workshop should incorporate these into a formal programme as follows:

- Stage 1 – Information
- Stage 2 – Structuring of objectives
- Stage 3 – Speculation
- Stage 4 – Evaluation
- Stage 5 – Development.

This is sometimes called the *job plan.* The detailed stages of the workshop are described below.

STAGE 1 **INFORMATION**

This stage focuses on the purpose of the project. It aims to establish a common understanding of the issues driving it and the key project constraints. An initial brief statement is made by the client or project sponsor describing the need for the new project. Stakeholders may want to challenge the assumptions in it and each participant is invited to make a short statement of their interpretation of the project objectives. It is important that each statement represents the speaker's own perception, rather than a reiteration of a previously agreed statement.

The key points from each statement are then discussed in full. To conclude, the facilitator produces a summary list of the agreed key objectives.

STAGE 2 **STRUCTURING OF OBJECTIVES**

Value for money cannot be achieved without a clear understanding of the project objectives and their relative importance. The objectives identified in Stage 1 should be structured into a *value hierarchy*, with one all-embracing objective at the top which captures the raison d'être for the entire project.

This higher order objective is then progressively broken down into lower order objectives using a 'means-ends' analysis, i.e. a lower order objective is considered to be the 'means-to-an-end' whilst a higher order objective is considered to be an end in itself. A typical value hierarchy is illustrated in Figure 7 overleaf and further guidance is provided in *Toolbox 4*. Judgement is often required to determine the extent to which the objectives should be sub-divided. For most projects two levels of breakdown should suffice.

It is important that the value hierarchy is constructed on the basis of *consensus* agreement. It is likely that several drafts will be needed before an acceptable end result is achieved.

FIGURE 7   TYPICAL VALUE HIERARCHY

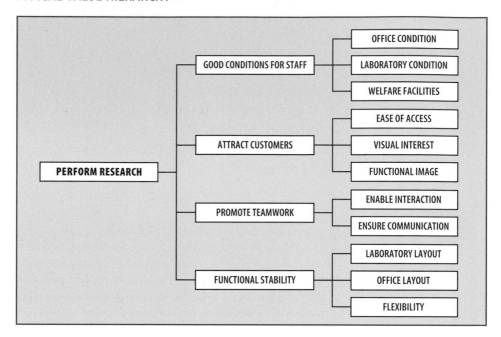

STAGE 3 **SPECULATION**

The purpose of this stage is to generate alternative ways of achieving the agreed objectives. It should not be taken for granted that a new construction project is necessarily the best way of doing this.

The value hierarchy identified in Stage 2 is used as the stimulus for a brainstorming exercise. Each objective is addressed in turn and the question asked 'How else might this be achieved?' It is important that no attempt is made to assess the feasibility of the ideas generated during this stage. The aim is to encourage a free flow of ideas. No criticism or ridicule should be permitted during brainstorming - the least likely ideas often provide the stepping stones to creative solutions which would not otherwise have been thought of. All team members must be actively encouraged to participate. The objective should be to generate as many ideas as possible in the time available; it is the *quantity* of ideas which is initially important rather than their quality. Further guidance on brainstorming is provided in *Toolbox 3.2.*

STAGE 4 **EVALUATION**

This stage marks a definite switch from creative to analytical thinking. The ideas produced in the previous stage should be combined and consolidated before their feasibility is evaluated. Those judged feasible are ranked in terms of perceived value and constant reference is made to the value hierarchy. The end product should be a small number of schemes, each of which would satisfy the required objectives. These may, or may not, include the original proposal for a construction project.

Further guidance on evaluation methods is provided in *Toolbox 4.2.*

> STAGE 5 **DEVELOPMENT**
>
> The task of developing and costing the preferred alternative schemes should, theoretically, also take place during the workshop. In practice it is often more convenient to do this outside the formal workshop. The final recommendation will probably depend upon the outcome of this detailed development work.
>
> The Value Management team must identify the extent of the follow-up work and agree a timescale. All participants should be fully aware of what has been agreed before the workshop ends.

DURATION AND STRUCTURE OF WORKSHOPS

4.3.3 The duration of the VM1 workshop will depend on the nature and scale of the project. Between one and two days should normally be sufficient for most projects.

Workshops can be structured in a variety of ways. While the VM1 workshop is well described in terms of the five key stages above, in practice they will rarely unfold in such a linear manner. In some cases it may be necessary to move between stages several times to ensure the involvement and commitment of all participants. In others, workshops can be split (for example, over a weekend) so that development work can be undertaken as part of Stage 5.

WHO TO INVOLVE

4.3.4 The participants in the VM1 workshop should include:
- the client (or project sponsor)
- senior representatives from all key project stakeholders
- the project manager and design consultants (if appointed)
- the value manager.

The team will be multi-disciplinary and should represent all interest groups, although its exact composition varies with each project.

RESULTS

7 *Toolbox* 3.3 provides further details of workshop reports.

4.3.5 Once the VM1 workshop has been completed, the facilitator presents a report to the client[7]. VM1 should result in a clear statement of project objectives, structured and ranked in a value hierarchy, and an initial set of design options for further evaluation and development.

The final outcome of VM1 could well question the need for the construction project initially envisaged, or may strongly reinforce it. In this case, the value hierarchy represents an important starting point for the design team in the subsequent development of the brief.

> **CASE STUDY – USING VALUE MANAGEMENT TO DEFINE OBJECTIVES**
>
> Value Management was needed to rationalise and define key objectives on a £800m railway upgrading project. From the outset it was clear that a range of financial, operational, technical and regulatory requirements would need to be met. An early challenge was to reconcile the views of the various stakeholders about what – and who – the project was for, and its impact on the existing railway.
>
> Value Management was undertaken in two stages. The first crystallised the project objectives. The second drew on these objectives to identify a range of possible options. By developing and costing the more promising options, and placing them within an overall value plan, it was possible to assess their relative value for money. Options could also be easily compared with similar rail projects elsewhere, both nationally and internationally. This helped ensure that the best value for money option was identified and developed to deliver the objectives set down in the business case and agreed by the key stakeholders.

4.4 VALUE MANAGEMENT AT FEASIBILITY (VM2)

OBJECTIVES

4.4.1 The purpose of a Value Management workshop at feasibility stage is to ensure that the client selects the outline design proposal which represents best value for money. It also addresses possible value improvements to the selected option.

The client's strategic brief continues to develop throughout feasibility and the designers normally respond to the client's requirements by producing a choice of outline designs. The VM2 workshop usually coincides with the decision point in the design process separating feasibility from scheme design. This is when the brief is frozen. The timing of the VM2 workshop is therefore critical: too late and the important decisions will already have been made; too early and there will be insufficient information available to make a judgement.

The specific objectives of Value Management at feasibility are to:
- verify that the project objectives are still valid
- ensure that the outline design proposal is chosen in accordance with the value for money criteria
- achieve a group consensus in favour of a single option
- secure value improvements in the chosen design option
- ensure that the decision-making process is accountable.

KEY ACTIVITIES

4.4.2 The plan for the VM2 workshop is typically:
- Stage 1 – Information
- Stage 2 – Structuring of objectives
- Stage 3 – Assignment of importance weights
- Stage 4 – Evaluation
- Stage 5 – Sensitivity analysis
- Stage 6 – Cost/value reconciliation
- Stage 7 – Value improvement.

The detailed stages of a VM2 workshop are as follows:

STAGE 1 **INFORMATION**

As with VM1, it is important to begin the VM2 study by focusing on the purpose of the project. The facilitator gives a short presentation outlining the objectives identified at VM1 together with any refinements which have since been made. There will almost always have been some changes, if only because the need for construction is now accepted. Participants are invited to comment on whether these objectives remain valid. The facilitator should then summarise the discussion and produce a revised list of project objectives.

This procedure is repeated with reference to the project *constraints*. It should not be taken for granted that all previous constraints are still valid and participants should discuss the basis of each one in turn. Finally, the facilitator summarises the surviving constraints and highlights those which may be relaxed.

STAGE 2 **STRUCTURING OF OBJECTIVES**

The value hierarchy is restructured in accordance with the revised objectives identified in Stage 1. However, unlike the VM1 workshop, the aim is to adopt the lower-order objectives as *attributes* for assessing different outline design proposals. In the example in Figure 7 on page 26, designs are assessed in terms of laboratory conditions, visual interest and so on. The number of attributes used should not be excessive and those which do not have a direct bearing on the choice of outline design should be eliminated. Whilst capital cost will be one of the factors used to assess the relative merits of design proposals, it should not usually be included as an attribute but should be considered as a separate issue at the end of the analysis.

STAGE 3 **ASSIGNMENT OF IMPORTANCE WEIGHTS**

It is clear that the attributes identified in Stage 2 will not all be of equal importance. One way of dealing with this is to assign weights to different values in the hierarchy. They should be initially allocated to the first level of breakdown, and then to each group of attributes in subsequent levels. A worked example is provided in *Toolbox 4.2.1*, Figure C.

The importance of weights must be agreed by group consensus.

Voting should be avoided.

STAGE 4 **EVALUATION**

Each design proposal can now be assessed against the weighted attributes. Proposals should be scored against each attribute giving weighted scores for each design proposal. The result is a comparative rating which represents the value of each design proposal, as measured against the predetermined attributes (see *Toolbox 4.2.2*, Figure D). Since the outcome depends on the importance weightings, their validity should be checked by examining how sensitive the results would be if they were changed.

STAGE 5 **SENSITIVITY ANALYSIS**

The purpose of this stage is to test how sensitive the outcome of the rating process is to changes in the key weights. Attention should be focused on those weightings or scores for which some participants may have expressed concern. If, by adjusting these weights the ranking order of the design proposals is affected, the weights should be considered more carefully. If such adjustments continue to support the ranking order, the participants can move on to the next stage with confidence.

It is important to recognise that this weighting procedure does not prescribe the 'correct' answer. It merely provides a formal framework for evaluation based on the professional expertise of the workshop members. Value for money can only be achieved by a rigorous evaluation process which tests the underlying assumptions. The more rigorous the process, the greater the confidence in the end result.

STAGE 6 **COST/VALUE RECONCILIATION**

The estimated capital and life-cycle cost of each proposal must now be considered. Comparisons may be made by calculating the ratio of each proposal's rating to its projected cost. Alternatively, the proposal with the highest rating could be chosen, provided that it does not exceed the project budget (see *Toolbox 4.2.2*). Small projected cost overspends can often be subsequently 'value-engineered out' without compromising the performance of the favoured option (see Part 5).

STAGE 7 **VALUE IMPROVEMENT**

Attention now focuses on how the chosen design option can be improved. The facilitator guides a discussion on areas of concern, which should be formally listed. A brainstorming session addresses each of the concerns in turn, thereby producing a list of ideas as to how they might be alleviated.

The ideas are then evaluated and a summary list is made of those meriting further investigation. Before closing, the facilitator ensures that all participants know what follow-up work they must do.

DURATION AND WHO TO INVOLVE

4.4.3 As is the case with VM1, the duration of the study is likely to depend on the size and complexity of the project in question. However, between one and two days should normally be sufficient (see also 4.3.3). The VM2 study workshop should involve the same participants as VM1, plus any new stakeholders who have since emerged (see also 4.3.4).

RESULTS

4.4.4 After the VM2 workshop the facilitator again presents a workshop report to the client. It should contain a clear recommendation about which design option offers the greatest value for money. This is critical, for it represents the point at which the brief becomes frozen. Changes to the brief after this point are likely to involve significant cost penalties, delay and disruption.

The workshop at VM2 should also produce a weighted list of project objectives and a list of suggestions for further improvement to the chosen design option.

4.5 PROCUREMENT STRATEGY AND VALUE MANAGEMENT

Value Management can be implemented regardless of how the works are to be procured. The initial study (VM1) at concept stage may precede the appointment of the design team and the choice of procurement method. Indeed, it could help formulate an appropriate method. Equally, by questioning why the project exists in the first place, it could recommend that construction work is not needed to achieve the objectives identified in the workshop.

When the initial study precedes the choice of procurement method, the procedures for Value Management at concept stage described in section 4.3 should be followed. If a particular method[8] has been chosen, it will influence the Value Management approach adopted.

[8] Different methods of construction procurement are explained in **Potter, M. (1995) *Planning to Build?*,** CIRIA.

TRADITIONAL

4.5.1 Under traditional methods, it is important to inform the design team prior to their appointment that Value Management will be used. The perception that Value Management is a crude cost reduction exercise, and the unhelpful defensive position which some in the team may adopt as a result, needs to be avoided. The design team should understand that effective Value Management is of great help to them in ensuring that they receive clear directions from the client. Designers respond positively to a process which reduces unnecessary changes to the brief and wasteful re-work. The success of Value Management can be measured by the extent to which all participants find it worthwhile. This is just as true for members of the design team as it is for members of the client organisation.

The design team must be adequately briefed before attending Value Management workshops. It is crucially important that they come prepared with the right level of information. The VM1 study should not be seen as replacing normal briefing activities and in the VM2 study each design option should be developed and costed to the same level of detail. The tendency to develop some favoured options to a greater extent than others should be avoided as this denies the client the opportunity to compare like with like.

DESIGN AND BUILD

4.5.2 Clients using Design and Build should be clear about their requirements and be able to communicate them effectively. Value Management can play a major role in making this process more effective.

There are two ways of using Value Management under this method. The first is when developing a robust brief ***before*** appointing a Design and Build contractor. Many clients engage an independent firm of designers to help develop the brief. In this case a VM1 study can help ensure that it is robust and that key stakeholders 'buy-in' to the statement of objectives.

The second is during the negotiations to finalise the appointment. Value Management at this stage would have two objectives:

- to ensure that the Design and Build contractor fully understands the client's requirements
- to provide a framework to use the contractor's buildability expertise to its full potential.

Cost savings at this stage can be shared between contractor and client. While savings identified after the contract is signed can be kept by the contractor, incentive clauses can be included in the contract (see 5.5.5, page 43) to share savings between both parties (see also 5.6.2, page 44).

Some Design and Build contractors offer to facilitate Value Management studies during the negotiations to award the contract. Clients considering this are encouraged to participate actively in such exercises for the benefit of both parties.

CONSTRUCTION MANAGEMENT

4.5.3 Value Management can build on the close collaboration between the designer and construction manager which exists under construction management. Clients can request their construction managers to include the facilitation of Value Management workshops within the package of services they require. Indeed, the provision of Value Management can be seen as an essential element in the wider design management strategy.

Construction management is often used on congested sites where new construction work has to proceed alongside existing operations. One of the client's key objectives may be to minimise disruption. In these circumstances, the relative value of different design options will depend on the associated temporary works and construction processes. The contribution of the construction manager is therefore vital during even the very early stages of the design process. The Value Management strategy described in this part of the guide will provide a basis for drawing together the different types of expertise necessary to identify valuable and innovative solutions.

BUILDING COST-EFFECTIVELY: VALUE ENGINEERING

PART FIVE

5.1 INTRODUCTION

WHAT DOES VALUE ENGINEERING DO?

5.1.1 This part of the guide covers the use of Value Engineering to achieve agreed objectives in the most effective way. It is relevant for all stages of projects where objectives are clear and well shared among the key stakeholders. It is also relevant for projects where objectives are ambiguous initially but are subsequently clarified and agreed. If objectives remain ambiguous and stakeholders are finding it difficult to agree, Value Management (as described in Part 4) is the more appropriate approach.

Value Engineering provides a structured approach to ensure that agreed project objectives are achieved effectively by:
- clarifying the functions required
- identifying possible ways (design options) of providing these functions
- evaluating the effectiveness of each option, in terms of its relative
  - cost, including life-cycle cost
  - buildability
  - performance.

This provides a sound basis for choosing effective options. In particular, it can:
- provide new insights into the functions required and the means of providing them
- eliminate unnecessary cost (including operational cost) which does not contribute to the required level of performance
- improve buildability
- improve performance and value.

Although the emphasis in current Value Engineering practice tends to be on eliminating unnecessary cost, it can also be used to improve other aspects of design options, such as project programme and safety (Value Engineering can form a valuable contribution to ensuring compliance with the Construction [Design and Management] Regulations 1994 [CDM][9]). Equally, it can be used to evaluate and improve less quantifiable aspects, such as overall design quality, aesthetics, user-friendliness and so on, though judgement will be required to appraise the benefits. Value Engineering does not simply seek the cheapest option, nor is it in conflict with excellence in design or construction. Rather, it spells out to clients the value of different options so that they can make informed decisions about which one offers the best value for money. This may involve spending more (rather than less) money to improve project performance and value.

[9] Health and Safety Executive (1994) provides a guide to the CDM regulations. See *Toolbox 6.*

> ## CASE STUDY – **USING VALUE ENGINEERING TO ENHANCE VALUE**
>
> A developer proposed to construct a speculative office building in London with a net lettable area of 4,500 square metres and at a cost of some £5m. Initial studies indicated that providing the necessary space within the site and cost constraints would be particularly challenging.
>
> The developer decided to use Value Engineering and employed a facilitator to work with the design team to find the most effective solution. The facilitator convened a workshop during which it became clear that the relationship between net lettable area and the size of the service cores was crucially important to the viability of the design. Although the designers had already evaluated this, the combined efforts of all parties working together creatively in a facilitated workshop environment identified a number of improvements to the outline design proposals. Potential improvements were also identified in the proposed wall cladding system, and these were subject to more detailed study outside the workshop.
>
> As a result of the workshop, the building cost 2% (£110,000) more than the original budget. However, the increase in benefits of some £3.4m (capitalised) more than paid for this.

RIGHT FIRST TIME

5.1.2 Value Engineering is primarily concerned about ensuring that correct design decisions are made first time. Although it can be used to review decisions which have already been taken (see 5.3.3), it is far more effective when used to ensure that design decisions are made in a systematic way. It can be highly effective when used alongside traditional cost control methods, rather than as a 'safety net' for correcting overspends. Planning to use it retrospectively would inevitably result in abortive work and have an adverse effect on the project programme.

VALUE ENGINEERING AND COST PLANNING

5.1.3 Some argue that there is little difference between traditional cost control and Value Engineering. Indeed, the best practice of construction cost control is similar in many respects to Value Engineering, but it is not always so powerful.

Traditional cost control during design development is usually based on a 'cost planning' system, whereby the design evolves within a predetermined cost for each main element. While this can help produce design 'to a cost', it is not explicitly concerned with **value**. Used creatively, of course, cost planning can produce valuable and cost-effective designs. However, when faced with cost over-runs there is a tendency to seek savings by downgrading specification which can reduce project performance and value.

5.2 THE PRINCIPLES OF VALUE ENGINEERING

There are several different techniques for effective Value Engineering, but the underlying principles are the same.

KEY REQUIREMENTS

5.2.1 With Value Engineering the design team, contractors, construction managers and specialist contractors should adopt a challenging and systematic approach to their work. They should always consider a range of options for each design element and work package before making a commitment to any one of them. Only by considering alternatives and assessing them against the agreed functional criteria can the client be assured of value for money.

To achieve the required results, a climate must be created which encourages multi-disciplinary co-operation. Effective design does not depend on any single specialism; it depends on the close collaboration of all. Close contact should be maintained between different design team specialists and every effort made to harness the expertise of specialist contractors for the benefit of the client (this is often achieved through the use of an incentive scheme - see 5.5.5).

Successful Value Engineering depends on:
- a multi-disciplinary approach
- developing a shared commitment to achieving project objectives
- developing the attitude which seeks to beat cost targets (see 5.2.4)
- encouraging responsible innovation.

THE IMPORTANCE OF 'FUNCTION'

5.2.2 The power of Value Engineering comes chiefly from its concern with function. While traditional cost control is mainly concerned with what the design elements are, Value Engineering is more concerned with what they actually do, i.e. their **function**. This approach considers cost in relation to function, recognising that there is a three-way relationship between function, cost and value. This relationship can be expressed visually as follows:

$$\textbf{VALUE} = \frac{\text{FUNCTION}}{\text{COST}}$$

Therefore, if both function and cost are reduced, there is no increase in value. Indeed, if function is reduced to a greater extent than cost, value decreases. Conversely, value can be improved by increasing cost, provided that the additional expenditure is justified by an improved functional performance.

Only by looking very carefully at what functions are required, and how different options will provide them, can clients decide how best to spend their money. A technique called function analysis helps do this and further guidance is provided in *Toolbox 4.3.*

ELIMINATING UNNECESSARY COST

5.2.3 The basic premise of Value Engineering is that a certain amount of unnecessary cost is inherent in every design. It is usually only possible to eliminate this by identifying another option which provides the same function at less cost. Unnecessary cost is the difference between the cost of an existing option and a better one. Because of the complexity of construction design, it is rare to find an option which is optimum in every respect. But it is important to seek improvements by considering the costs and benefits of alternatives before making a choice.

Specific causes of unnecessary cost include the following:
- Cost of unnecessary attributes
    - the cost of attributes which provide no useful function
- Cost of unnecessary specification
    - the cost due to needlessly expensive materials and/or components specified
- Unnecessary cost of poor buildability
    - the cost due to a failure to consider construction implications of design choices
- Unnecessary life-cycle cost
    - the cost due to a failure to consider future operational costs of design choices
- Unnecessary opportunity cost
    - the cost of losing potential revenue. Failing to generate revenue is just as important as wasteful expenditure.

COST TARGETS FOR VALUE ENGINEERING

5.2.4 Project cost targets are an important component of Value Engineering. They should not be set simply to be achieved, but to be beaten. The aim should always be to deliver each element of the design at lower cost. The cost-effectiveness of different options can be measured against the cost targets established for each main element.

Ideally every design decision should be subject to Value Engineering, but *__80% of cost is often contained in 20% of the design decisions.__* Most effort should, therefore, be concentrated on those decisions which have the greatest impact on cost.

Typical examples of the high cost elements within different types of construction projects are presented in *Toolbox 5*. Other topics may, of course, warrant specific investigation. Presenting a cost plan in graphical form can often help focus designers' attention towards high cost areas (see, for example, *Toolbox 4.3*, Figure H).

5.3 USING VALUE ENGINEERING

RESPONSIBILITIES **5.3.1** The responsibility for Value Engineering (and the methods used) depend on how the project is procured. But a number of key principles apply in all cases. Specific guidance for different procurement approaches is given in Part 5.6.

The initial responsibility for creating an atmosphere which encourages the multi-disciplinary collaboration needed for effective Value Engineering rests firmly with the client's project manager, who should lead by example. The importance of Value Engineering should be continually emphasised to all in the construction team who should adopt a challenging and systematic approach to their work in a spirit of teamwork and co-operation.

Such co-operation implies a shared commitment to work towards the client's objectives and **shared responsibility** for delivering an effective project. This is quite different to the adversarial climate which often prevails in construction, where different members of the team blame each other for difficulties that arise, rather than seeking effective solutions. The idea of shared responsibility is fundamental to getting it right first time. The ability of consultants and contractors to work in this way should be assessed prior to their appointment

66 *To embed Value Engineering into your organisations will take time and patience. If it is done too hastily, then the initiative can flounder on the disbelievers. If it is done too slowly, then you are in danger of losing momentum and not achieving your goals. I would suggest that if your marketplace is seeing the same competitive pressures as mine, that you persevere with it for your own long-term benefits.* 99

Source: Presentation by Neil Hegarty, General Manager of Technology and Quality, BOC Process Plants, to a conference organised by SBIM Ltd, September 1993.

VALUE ENGINEERING AWARENESS **5.3.2** It is particularly important for external consultants to be fully briefed on appointment regarding the intention to use Value Engineering. Indeed, their understanding and experience of Value Engineering may be among the criteria by which they are selected.

As with Value Management in the early project stages, it is important to run Value Engineering 'awareness sessions' for participants who are unfamiliar with the approach. Part 4.2.5, page 23, provides further details.

VALUE ENGINEERING REVIEWS **5.3.3** If during design and/or construction, it becomes apparent that the budgeted cost is going to be exceeded, or if the client is unhappy about a particular element of the design, a formal Value Engineering workshop may be needed to review the chosen design solution. **This should be the exception rather than the rule, especially when a planned approach to Value Management is used throughout the project.** Although a formal Value Engineering review can often produce powerful results, there is a danger in over-using this approach. If the construction team expect reviews to take place as a matter of course, the importance of making cost-effective decisions in the first instance is likely to be under-emphasised.

CASE STUDY – USING VALUE ENGINEERING RETROSPECTIVELY TO REDUCE COSTS

A manufacturing company commissioned the design of a new factory.
It became apparent at the detailed design stage that the projected costs (£60m) exceeded the project budget. Because of the need for a swift and in-depth review, the company decided to undertake a value engineering exercise.
An external facilitator was appointed to help the design team review the design.

The factory design was a linear building, enclosing seven straight, single-speed, production lines. The study helped to re-engineer the layout into five multi-speed lines. These lines offered improved performance over the seven single-speed lines and allowed the product to be programmed down more than one line. Further, by arranging some lines at 90° to others, the overall length of the factory enclosure was reduced. This significantly altered the site layout and enabled some ancillary storage to be planned more efficiently. Altogether, some 4,000m^2 of the originally proposed building was no longer required.

The net result of the study was to reduce the projected cost by 19% from £60m to £49m, and to improve the performance of the facility.

5.4 DESIGN OPTION APPRAISAL

APPRAISAL PRINCIPLES

5.4.1 Generating and appraising alternative design options is central to Value Engineering. If value for money is to be achieved, a rigorous method for design option appraisal must be adopted.

Where different options provide the same functional performance, then the appraisal need only concentrate on their relative costs and their implications for project duration and ease of design and construction. Options will not always offer the same functional performance and a more thorough appraisal of their relative costs and benefits may be needed. Of course, the client who is satisfied with a clearly defined minimum level of performance may consider any benefits above this to be excessive. But if improved performance might increase value for money, it may be necessary to apply judgement to appraise the benefits.

APPRAISAL METHODS: LIFE-CYCLE COSTING

5.4.2 Life-cycle costing is a commonly used appraisal technique. It provides a basis for examining and comparing the costs and benefits of an option over its life. It does this by 'discounting' all costs and revenues (both now and in the future) of each design option to convert them to a common base (the Net Present Value – NPV). Options can then be compared on the basis of NPV to help identify which offers best value for money. The benefits of options offering improved performance can be included in the NPV comparison if expressed in financial terms. The following must be established for a life-cycle costing appraisal:

- the discount rate
- the 'time horizon' (this will normally be the design life of the completed project, or the duration of the client's interest in it)
- the cost and benefits of differences in project duration and ease of design and construction ('buildability', see 5.4.3)
- the costs and benefits of other desired improvements.

Different options should be compared over a similar time scale. Alternative elements or components having different life expectancies are normally compared over the total life of the completed project. Further guidance on the calculation of NPV and other life-cycle costing methods is provided in *Toolbox 4.4*.

BUILDABILITY APPRAISAL

5.4.3 The capital cost of a design option depends on the ease with which it can be constructed ('buildability') and on the materials specified. For example, construction costs vary according to the extent of straightforward and sequential construction, the number of interfaces between different construction operations and the need for temporary works. The buildability of each option must therefore be evaluated.

This normally includes a comparison of costs and benefits of each of the following:
- the construction programme, including delivery times for materials and components
- the balance between site production and off-site prefabrication
- the logistics of site operations, including the number of interfaces between trades
- the need for temporary works
- the extent to which existing operations, including those adjacent to the site, will be disrupted.

5.5 VALUE ENGINEERING APPROACHES

The main Value Engineering approaches covered here are:
- concurrent studies
- workshops
- contractor's change proposals.

CONCURRENT STUDIES

5.5.1 **CONCURRENT STUDIES** consist of short, regular brainstorming sessions/meetings between the value manager and the construction team. They are used throughout the design and construction phases to review key design decisions and are very effective in promoting a continuous Value Engineering philosophy or culture. They are particularly suited to management-type contracts where the design of work packages is staggered and overlaps with construction. Although they can be used as an alternative to Value Engineering workshops (see 5.5.2), there is a danger of evaluating individual elements of the design in isolation and losing the overall picture. Concurrent studies can be a very effective means of identifying cost savings during detailed design. They are less disruptive than a one or two-day Value Engineering workshop, although they can be disruptive and costly if held too frequently. They should therefore concentrate on key design elements only.

THE VALUE ENGINEERING WORKSHOP AND JOB PLAN

5.5.2 The established procedure for a formal Value Engineering **WORKSHOP** is often referred to as the 'job plan'. It can be used at all relevant stages of construction projects (see Part 3, Figure 3) and typically consists of the following five distinct stages:

STAGE 1 **INFORMATION**

The first step is to identify areas within the developing design which offer the greatest potential for eliminating unnecessary cost and/or improving cost-effectiveness. This is followed by information gathering which focuses attention on the nature of the options being studied. The aim is to identify and examine the functions that are needed of the main design elements, starting with the primary function, and asking 'what does it do?'. A technique called function analysis can be used for this. Further details are given in *Toolbox 4.3*.

STAGE 2 **SPECULATION**

The second stage of the job plan concerns generating ideas about other ways of providing the primary function. Creative thinking techniques (such as brainstorming) are used in order to produce as many ideas as possible. Further guidance is provided in *Toolbox 3.2*.

STAGE 3 **EVALUATION**

Once the speculation process has been exhausted, the ideas generated should be evaluated. They are examined in terms of their feasibility, cost and, if appropriate, improved functional performance. Initially the key criterion for assessment is whether the functions identified in the Information Stage are met. Often ideas need to be combined and consolidated in order to meet the functional requirements. This process is likely to narrow the number of feasible ideas down to, perhaps, five or six, though more may be possible.

STAGE 4 **DEVELOPMENT**

A limited number of ideas are carried forward for further development. They are then designed in more detail and carefully examined for cost, performance and buildability. Each option should be appraised on a life-cycle costing basis. The Value Engineering team (see 5.5.3) may well confer with outside specialists during this phase, and then make its recommendations.

STAGE 5 **RECOMMENDATION**

The final stage of the job plan is probably the most important. Its purpose is to decide which option represents best value for money and to ensure that this recommendation is implemented.

This procedure should be applied separately to each of the design elements under consideration. Initially effort should, however, be limited to the first three stages of the job plan; the detailed development work and recommendation stages may then take place outside the formal workshop, with the final recommendation being agreed at a subsequent design team meeting. Whilst a certain degree of free-ranging discussion is both inevitable and beneficial, it is important that the overall structure is followed. A truly systematic review cannot be carried out without a rigorous framework.

COMPOSITION OF THE VALUE ENGINEERING TEAM

5.5.3 An issue of crucial importance is who is in the Value Engineering team. The team should be multi-disciplinary, made up of people with as wide a range of expertise as possible, including buildability, maintenance management, cost modelling, architectural design and engineering design. It is also often beneficial to include a non-construction professional to act as 'devil's advocate'.

Some practitioners argue that the Value Engineering team should consist entirely of 'fresh faces' previously uninvolved in the design. Such a group has the advantage of doing the workshop with no preconceptions. They can reassess the basic assumptions made by the existing design team. However, the danger is that such a group may come into conflict with the existing design team (see below). It is often more effective therefore to use the existing design team, supplemented as necessary by 'external' expertise. This team is led and guided by a facilitator, who helps team members carry out their own Value Engineering.

> In the USA the classic Value Engineering exercise is a **40-HOUR WORKSHOP** attended by the value manager and an independent design team. The findings are reported to the client and project manager for further action/implementation. While an external design team has the advantages of providing a fresh and critical approach and an independent review, in the UK the disadvantages are generally believed to outweigh them. These include:
> – conflict with the existing design team
> – loss of time while the external team becomes familiar with the project
> – the additional cost of a second design team
> – delay and disruption to the design process during the review
> Also, the external team may feel 'obliged' to identify cost savings to 'justify' their fee.

While the workshop can be chaired by someone from within the client or design teams, an external facilitator is often more effective. The precise make-up of the team varies with the works procurement method used, and the problem being addressed. On a typical building project it might comprise the following:

> **THE VALUE ENGINEERING TEAM**
>
> | | |
> |---|---|
> | • Facilitator | • Services Engineer |
> | • Client representative/ key stakeholders as appropriate | • Quantity Surveyor |
> | | • Construction Engineering Advisor |
> | • Project Manager | • General Contractor |
> | • Architect | • Specialist Contractor |
> | • Structural Engineer | • Facilities Manager |

DURATION AND TIMING OF WORKSHOPS

5.5.4 Value Engineering workshops should take place when the key design decisions are being made and all parties should agree workshop dates in advance.

The duration of workshops will depend on the complexity of the project and the number of design elements to be considered. As a general guide, between one and three days should normally be sufficient, though workshops can last up to five days and longer.

Workshops can be structured in a variety of ways. For example, much of the detailed development work can take place outside formal workshops which can be split (for example, over a weekend) to accommodate this.

CONTRACTOR'S CHANGE PROPOSALS

5.5.5 **CONTRACTOR'S CHANGE PROPOSALS** are design and/or construction changes proposed by the contractor, either at tender stage or after the contract has been awarded, and primarily aimed at reducing costs. They are a useful way of drawing on the contractor's knowledge and experience to improve buildability and reduce costs. These change proposals are often linked to an incentive scheme for the contractor and, if this is the case, clients should ensure that quality and maintainability do not suffer. They should also be aware that the design team may need to check the feasibility of change proposals, and this may lead to delays. Clients should therefore carefully weigh up the potential cost savings/improvements arising from change proposals against the likely costs and programme implications of implementing them.

IMPLEMENTATION AND FEEDBACK

5.5.6 Value Engineering exercises normally result in a range of recommendations for design modifications and/or further design development. For workshops, these are summarised in a report produced by the facilitator (see *Toolbox 3.3*) within three to four days. The client must consider carefully those recommendations which may require additional expenditure or significant changes to the project brief before they are implemented. Also, clients who wish to adopt a long-term Value Management philosophy and learn from past experiences should obtain feedback on the success/failure of each approach. These issues are covered in Part 6.

Successfully implementing recommendations from Value Engineering exercises requires:
- an agreed timescale for implementation
- adequate briefing for those responsible for implementation.

DESIGN LIABILITY/ RESPONSIBILITY

5.5.7 Value Engineering is essentially a 'no fault/no blame' process. Recommendations must be agreed by the design team if they are to be implemented. Instructions issued by the client or project manager to the design team as a result of Value Engineering should be no different to other design instructions issued on the project. The design team is responsible for carrying out any changes to the design as a result of these instructions and will be liable for them.

5.6 PROCUREMENT STRATEGY AND VALUE ENGINEERING

The Value Engineering techniques used and individuals involved depends on how the project works are procured. The approach can be effective, however, regardless of the procurement method. The following paragraphs highlight the key implications for Value Engineering of the more common forms of procurement.

TRADITIONAL

5.6.1 On projects with a 'traditional' design team and a general contractor, the responsibility for effective Value Engineering rests firmly with the client's project manager. In addition to any formal studies which may be planned, there is a continuing need to ensure that the designers have considered a wide range of alternatives before adopting specific solutions. This is best achieved if the project manager, designer and cost consultant all work closely together.

Value Engineering reviews are particularly useful if the budgeted cost is likely to be exceeded (see also 5.3.3). They then provide a forum for a wider range of expertise than would normally be available to the design team. Reviews are more common with traditional procurement, but they are still the exception to a planned series of Value Engineering studies.

Because traditional design is separate from construction, the design team may not have sufficient expertise to consider fully the buildability of different design options. This can adversely affect construction cost and programme. A formal Value Engineering workshop to address the construction implications of specific design options may therefore be beneficial. Improved communication within the construction team can bring additional benefits.

DESIGN AND BUILD

5.6.2 With design and build, there may appear to be little benefit to clients in using Value Engineering once the contract has been awarded.

10 For a discussion of payment mechanisms for building contracts, see **Potter, M. (1995)** *Planning to Build?*, CIRIA.

Under 'lump sum' contracts, Value Engineering may be initiated by contractors for their own benefit, rather than to pass savings on to clients. Incentive clauses (see also 5.5.5) based on a formula for sharing the savings identified can, however, be included in construction contracts, most readily in target price or target cost reimbursable contracts[10]. In this case, the contractor's cost accounting and payment arrangements will need to be transparent so that savings can be clearly identified and agreed.

When using design and build, value for money may be obtained by examining the quality and performance of alternative contractors' designs. These must be comparable if cost is to be used as the sole criterion of selection.

CONSTRUCTION MANAGEMENT

5.6.3 The implementation of Value Engineering is fundamental to the construction manager's role. Indeed, construction managers may be selected partly on the basis of their Value Engineering experience and track record and they can help plan a series of Value Engineering studies. In addition to these formal studies, they must work closely with designers 'around the drawing board' to ensure that effective designs are developed. They should also make full use of the expertise of specialist contractors.

The construction manager is expected to justify, with supporting evidence, key decisions taken during design development. The client's project manager can ask for a Value Engineering report on a particular building element where it is felt justified to do so. This report should contain a full description of alternative design options considered and the results of the appraisal.

VALUE MANAGEMENT IN A CLIENT ORGANISATION

PART SIX

6.1 IMPLEMENTING A VALUE MANAGEMENT SYSTEM

Value Management can be applied to an organisation's ongoing capital programme – and indeed, to a wide range of allocation and management problems – as well as on a project-by-project basis. This section provides initial guidance for client organisations who build regularly and wish to achieve value for money on all their projects.

While regular clients can appoint Value Management consultants for individual projects, they may also benefit in the long term by developing their own in-house expertise. This expertise needs to be developed progressively to a planned strategy. The key elements of such a strategy are described in the following paragraphs 6.1.1 to 6.1.6.

APPOINT A CHAMPION

6.1.1 When seeking to implement Value Management throughout an organisation it is important to appoint a 'champion' at an early stage. This appointment should ideally be made from within. A full understanding of the organisation's procedures and culture is crucial. Given that Value Management is likely to cut across existing vested interests and responsibilities, it is also important that the person appointed is senior and has well-developed interpersonal skills. Furthermore, the champion should be independent of existing power structures, and must not therefore be associated too strongly with any one department. He or she will act as project manager for the implementation process, liaising with consultants where necessary and acting as sponsor when consultants are appointed.

Once the Value Management system has been successfully put in place, it may be appropriate for the champion to hand over to an appointed 'Head of Value Management'. People who are good at implementing new initiatives are not always good at managing them in the longer term. It is crucially important that whoever is appointed to manage the implementation of Value Management has full board-level support. Effective Value Management programmes are best implemented from the top down.

DEVELOP AND DOCUMENT VALUE MANAGEMENT GUIDELINES

6.1.2 The strategy (and terminology) adopted should be summarised in a guidance document or manual. It is particularly important to relate the sequence of Value Management studies to the client's existing procedures for capital expenditure approval. Value Management should enhance existing procedures, not replace them.

The guidance document helps ensure that there is a common understanding of Value Management and its benefits. The client's champion must maintain an active involvement even if a consultant is appointed to help draft the document. It is also important to recognise that consultants experienced at facilitating Value Management workshops may not be best placed to advise on implementing a Value Management system. Clients should use consultants who have experience in this type of work. They will probably also be commissioned to run a series of awareness seminars and a facilitation training programme.

The guidance document should be written after wide consultation throughout the organisation, and all relevant departments should be actively involved. The final document should be a surprise to no one. It is only through the active participation of all interested parties that commitment and ownership will be fostered. The document must be well structured and readable.

AWARENESS SEMINARS

6.1.3 In addition to the guidance document, a number of Value Management awareness seminars should be held throughout the organisation. They can play a key role in establishing the credibility of Value Management and ensuring that there is widespread 'buy-in'. The seminars should be jointly presented by the in-house champion and the appointed consultant, and each one can be introduced by a senior manager. They should be approximately one hour in length and should be designed for those who will take part in Value Management workshops.

The seminars should therefore concentrate on the following issues:
- the benefits of Value Management
- Value Management success stories elsewhere
- the relationship of Value Management to existing procedures
- the role of workshop participants.

The seminars play an important role in correcting any misconceptions about Value Management and in making sure that the key issues are understood throughout the organisation. Each seminar should allow sufficient time for questions and discussion. Questions and perceived problems must be dealt with openly.

> 66 *It is important that departments and project sponsors are committed to the introduction and implementation of VM, being well informed about what to expect and who to involve. Only then can the techniques be effectively introduced and the necessary resources and support provided.* 99
>
> Source: **HM Treasury (1996)** *Value Management,*
> *Central Unit on Procurement, Guidance No 54,* **January.**

PILOT WORKSHOPS

6.1.4 Nothing helps to establish the credibility of Value Management more than successful results. Clients should, therefore, commission a number of pilot Value Management workshops during the early stages of implementation. Three would normally be sufficient. They should be carried out on high profile projects to ensure maximum impact.

The pilot workshops should take place when the guidance document has been published and the awareness seminars have been held. Using a different consultant for each workshop has the advantage of being able to see different approaches in action.

FACILITATOR TRAINING PROGRAMME

6.1.5 Training in facilitation is required for suitable in-house personnel. Participants should be drawn from a wide cross-section of backgrounds - the most effective facilitators are not necessarily construction professionals. The external consultant involved in drafting the guidance document will usually do most of the facilitator training. Once again, it is important that the client looks for a track-record in similar work.

The training programme should start with an intensive 2–3 day course providing hands-on experience in using Value Management techniques. The course organiser should provide each participant with a detailed set of course notes, which must complement the guidance document.

The initial training course can be developed by a programme of 'mentoring'. Additional trainees can observe an experienced facilitator in action – the pilot workshops (see 6.1.4) can be used for this. They could then work alongside an experienced facilitator, before undertaking facilitation themselves, with a skilled facilitator giving them support. It should then be possible to judge whether they are capable of working on their own.

There is an inevitable drop-out rate in most professional training programmes. An initial batch of 20 trainees might produce 8–10 active facilitators who can undertake their Value Management duties either full- or part-time.

ESTABLISH VALUE MANAGEMENT UNIT

6.1.6 Co-ordinating an organisation's Value Management procedures requires an administrative centre. This is initially headed by the Value Management champion who later hands over to a full-time Head of Value Management. Systems need to be set up so that the Value Management Unit knows about new projects and progress on existing projects, workshops can be organised and facilitators appointed. It must also be possible to collate and distribute workshop reports within one week of each workshop.

An important principle is that facilitators should not lead Value Management on a project in which they are also stakeholders. For example, project managers and designers should not facilitate workshops on their own projects.

6.2 MAINTAINING A VALUE MANAGEMENT SYSTEM

STRATEGIC IMPORTANCE OF FEEDBACK

6.2.1 Value Management can contribute to the continuous improvement of how clients procure construction. This includes the activities of briefing, designing, constructing, refurbishing and post-occupancy evaluation. Continuous improvement depends on feedback to clients about the success and failure of different approaches. While some benefits may become apparent almost immediately following implementation, others may not be proven until the project has been completed, handed over and is well into occupation/use.

POST-PROJECT EVALUATION

6.2.2 Post-project evaluation is one form of feedback which can demonstrate how well the objectives identified during Value Management were achieved. Its main aim is for clients and practitioners and, indeed, the entire project team to learn from the process and improve future performance.

Post-project evaluations are often best carried out in a workshop, ideally with the same stakeholders who were involved in the Value Management studies. A commitment to post-project evaluation fosters long-term learning in the procurement of construction. It highlights how the needs of stakeholders change over time and can also identify the need for new construction. The results of post-project evaluation can often be used to define the design objectives for new projects.

CONTINUED
INNOVATION

6.2.3 Value Management is only of value if it stimulates people to innovate and think critically about their decisions and procedures. In the long term, clients need to be aware of the dangers of 'workshop fatigue'. Participants become tired of just one approach. Clients must keep up-to-date with new ideas and continue to develop their approach to Value Management.

Even where a client has developed in-house expertise, it is often useful to use consultants on occasional projects in order to keep up-to-date with new ideas. In-house personnel should continue to attend training programmes and keep abreast of developments in Value Management practice and research. Clients' Value Management systems likewise need continuous innovation and may require re-structuring from time to time. If Value Management becomes 'just another routine' it will not achieve its objectives.

THE VALUE MANAGEMENT TOOLBOX

APPENDICES

TOOLBOX 1
GLOSSARY OF TERMS

AWARENESS SEMINARS
Seminars aimed at making stakeholders and other workshop participants aware of Value Management, its objectives, advantages and procedures.

BUILDABILITY APPRAISAL
Assessment of design solutions on the basis of construction programme, number of interfaces between construction trades, associated temporary works and extent of disruption to existing operations, all aimed at establishing the ease with which designs can be constructed.

CONCURRENT STUDIES
Value Engineering approach consisting of short regular workshops or brainstorming sessions between the Value Manager and the design team aimed at reviewing key design decisions.

CONTRACTOR'S CHANGE PROPOSALS
Cost reduction method consisting of post-tender design and/or construction changes proposed by the contractor. It is frequently linked to incentive schemes which reward the contractor for savings achieved.

COST PLANNING
Costing technique aimed at ensuring that designs and projects are cost-effective by checking that each element of the project or design evolves within a predetermined cost. Not explicitly concerned with value.

FACILITATOR
An expert in Value Management and Value Engineering who also runs workshops.

FAST (FUNCTION ANALYSIS SYSTEM TECHNIQUE) DIAGRAM
A diagrammatic representation of function which breaks down the identified basic function into its constituent secondary sub-functions. The diagram should structure the sub-functions into logically interrelated groups.

FUNCTION ANALYSIS
Technique designed to help understand the purpose of a component or process in terms of its underlying function.

IMPORTANCE WEIGHTING
Assignment of arithmetic weights to different values/attributes in a value hierarchy according to attributes' relative importance.

JOB PLAN
A logical and sequential approach to problem solving used as the basis of a programme for workshops, incorporating the key stages of problem identification and appraisal, identification and appraisal of possible alternatives and selection of preferred approach.

LIFE-CYCLE COSTS
Assessment of the costs of an asset over its lifetime (including initial capital costs, replacement costs, maintenance and repair costs). All anticipated future costs are converted to a common base by discounting them to a net present value.

MATRIX ANALYSIS
A technique for the evaluation of various options wherein scores are awarded for each option against key criteria. These scores are then multiplied by the appropriate *importance weighting* and the total weighted scores for each option are compared to identify which offers best value for money.

POST-PROJECT EVALUATION
Procedure following the completion of a project (frequently involving a workshop) to obtain feedback on the success/failure of all aspects of project performance. The aim is to learn from past experience and improve future performance.

RISK MANAGEMENT
Systematic approach to identify, assess, control and manage risk on projects in order to minimise potential damage.

STAKEHOLDERS
Investors, end users and others (i.e. all key interest groups) with a real interest in - and the power to influence - the project outcome.

VALUE ENGINEERING
A systematic approach to delivering the required functions at lowest cost without detriment to quality, performance and reliability.

VALUE FOR MONEY
Optimum combination of whole life cost and quality to meet the client's requirement.

VALUE HIERARCHY
Breakdown of project objectives according to their perceived value, from a primary all-embracing objective, which captures why the project exists, to successive lower level sub-objectives.

VALUE MANAGEMENT

A structured approach to defining what value means to a client in meeting a perceived need by clearly defining and agreeing project objectives and how they can be achieved.

VALUE MANAGEMENT UNIT

Administrative centre which co-ordinates an organisation's Value Management procedures. Tasks include organisation of workshops, appointment of facilitators and collation and distribution of reports and other material.

VALUE MANAGER

Expert on Value Management and Value Engineering (in-house or from an external firm) who can assist in developing and implementing a Value Management plan and organising Value Management and Value Engineering studies.

WORKSHOP

Meeting usually held at one of several 'pinch points' of a project, where key stakeholders and project participants work together to find the best value for money solution to a particular problem or situation.

TOOLBOX 2 APPOINTMENT OF A VALUE MANAGER: KEY SELECTION CRITERIA

Clients should appoint a ***value manager*** to assist them in developing and implementing a Value Management plan and to organise all aspects of Value Management studies.

Generally, value managers should have a thorough knowledge and experience of Value Management and its associated activities and techniques. They should be competent in a range of management and interpersonal skills (see box below).

SKILLS REQUIRED OF A VALUE MANAGER

The value manager should be able to:

- analyse complex problems
- seek innovative solutions
- challenge assumptions about needs and approaches
- motivate project participants towards achieving the project objectives
- communicate with both technical and lay project team members
- organise and facilitate workshops and brainstorming sessions
- provide authoritative leadership.

For construction projects, it is also useful for value managers to have a good understanding of the construction process.

If value managers are to also act as ***facilitators,*** they should have previous experience in facilitating workshops at key stages in project development.

SKILLS REQUIRED OF A FACILITATOR

The facilitator should be able to:

- ***guide*** team members through the Value Management methodology and ***act as a catalyst*** in the search of creative solutions. The facilitator should not merely be seen as a 'visiting expert'
- encourage team members to contribute their ideas
- foster the active participation of ***all*** workshop participants, e.g. by ensuring that strong personalities do not dominate the workshop
- manage time effectively and keep to the workshop programme
- maintain a clear view of the key issues at all times and not become 'bogged down' in detail.

It is usually an advantage for the facilitator to have an outgoing and dynamic personality.

TOOLBOX 3 WORKSHOP TOOLBOX

3.1 WORKSHOP ARRANGEMENTS

There are a number of preliminary activities which are crucial to the success of workshops.

PRIOR ARRANGEMENTS

> All workshop participants should be given:
> - sufficient notice of the date and venue of the workshop
> - the appropriate workshop documentation (e.g. the briefing document - covering the purpose, tasks and timing of the workshop - and a detailed agenda)
> - a statement of the purpose of Value Management and of what is expected of participants at the workshop.

The workshop documentation should be received in good time, i.e. at least one week prior to the workshop. It is important to provide participants with the *most relevant* information only and not to overload them.

WORKSHOP DOCUMENTATION

> - *For a workshop at concept stage (VM1)* a briefing document would typically include the client's initial statement of need and any information considered to be relevant to the project, such as:
>   - provisional information on cost limits/targets
>   - site details
>   - initial drawings and specifications
>   - details on main risks and constraints
>   - general background data supporting the need for the project.
>
> - *For a workshop at feasibility stage (VM2)* the documentation should typically include:
>   - the feasibility study and investment appraisal
>   - the initial statement of project objectives and constraints
>   - the outline programme and outline cost report
>   - the outline drawings and specifications
>   - life-cycle costs for design options (if available)
>   - details of any procurement proposals.
>
> - *For Value Engineering (VE) workshops (at scheme design or detailed design stages)* the documentation should typically include:
>   - the project brief
>   - the latest programme and cost report
>   - the scheme and/or detailed drawings and specifications
>   - life-cycle costs for design options.

The workshop **location and type of venue** should be carefully chosen to ensure that participants can work in comfort and free from interruption.

WORKSHOP VENUE

> The workshop venue should be:
>
> - **comfortable** for participants, as workshops can be very intensive:
>   - it should be well ventilated
>   - there should be a large conference table providing enough space for each participant
>   - provision should be made for regular refreshments (the facilitator should be responsive to the mood of the group and introduce breaks as and when required).
>
> - **well equipped:**
>   - it should be supplied with flipcharts, coloured marker pens and other visual display material
>   - there should be plenty of wall space to display individual flipchart pages summarising the results of each stage.
>
> - **neutral,** i.e. it should not be the normal place of work of workshop participants. In this way participants are isolated from their normal activities and interruptions are minimised.

3.2   BRAINSTORMING GUIDELINES

In group sessions there is frequently a danger that individual members will tend to conform to dominant opinions, which can stifle the creative development of the whole group. This is where techniques to improve creativity and decision making, such as brainstorming, can be particularly useful.

PURPOSE

The purpose of brainstorming is to **generate a wide range of ideas** from each participant, with the aim of solving the problem under investigation. Frequently the ideas which do not appear so promising initially turn out to be the best ones. It is the amount of ideas rather than their quality which is initially important and the principal rule of brainstorming is that ideas are not judged nor criticised when they arise; they are evaluated at a later stage in the workshop. This encourages the more reticent participants to contribute their ideas.

KEY PRINCIPLES

> **The key principles of brainstorming include:**
> - ideas should be generated, not evaluated
> - as many ideas as possible should be generated - some good ideas may arise spontaneously whereas others may originate from combining or building upon other ideas (but not from criticising them)
> - all ideas should be welcomed, no matter how radical
> - all ideas should be documented
> - innovation should be encouraged
> - participants should be encouraged to combine and improve each others' ideas, without criticising or evaluating them.

SUCCESS FACTORS

> *The success of brainstorming sessions is heavily dependent on the ability of the facilitator to establish and maintain an appropriate 'climate':*
>
> - ideas can sometimes 'run dry' and the facilitator's interpersonal skills should be used to create new stimulus and momentum
> - innovative ideas can sometimes become blocked - the facilitator should be able to recognise and overcome this problem
> - the facilitator should always encourage an open forum where each participant's contribution is equally valued.

PROCEDURES   The ***ideas generated*** are all recorded on a flipchart and are ***edited*** to remove obvious repetition. The remaining ideas are then ***categorised*** and, finally, ***evaluated*** and ***ranked*** on the basis of consensus. Suggestions for the implementation of the best ideas can then be made.

OTHER TECHNIQUES

> Apart from brainstorming, there are also other techniques which can help to improve group decision making. Two of these are outlined below:
>
> - the **NOMINAL GROUP TECHNIQUE** is very similar to brainstorming, but it is more formal in that *no communication* at all between participants is allowed during the initial phase of idea generation (as interaction between participants can inhibit creativity) and there is also little discussion at later stages. The procedure is as follows:
>   - ideas are first generated individually and written down by each participant
>   - they are then presented to the group, one at a time, without any discussion taking place, and are all recorded on a flipchart
>   - participants then explain their ideas and comment on the ideas of others
>   - each participant then ranks the ideas individually
>   - finally, the rankings are combined/pooled to help reach a consensus or common understanding.
>
> - the **'DELPHI METHOD'** using *written questionnaires* can be advantageous in cases where participants find it difficult to attend a workshop. The process is *anonymous* and involves the following stages:
>   - respondents reply to an initial questionnaire
>   - the ideas generated are then collated by a monitoring team
>   - the collated ideas are then fed back to respondents so that they can compare their opinions with those of other participants and adapt their initial ideas
>   - the respondents then send their new/updated responses
>   - the new/updated responses are then analysed by the monitoring team to establish if there is any consensus or if divergent opinions still remain
>   - if there is still divergence more iterations can be introduced until consensus is finally achieved.

ACHIEVING CONSENSUS Achieving consensus at workshops can sometimes be difficult. It is important to note that *real consensus is not achieved by covering over latent disagreements but by bringing them into the open* and finding an acceptable solution.

> The following general 'rules' on achieving consensus should be followed:
> - differences of opinion should not be taken personally
> - differences of opinion should not be seen as an impediment, but as the basis of an improved solution
> - participants should never change their mind just to avoid conflict; they should only do it if they are genuinely convinced
> - techniques to avoid conflict, e.g. majority vote and coin flipping should be avoided.

In cases where a deadlock is reached, the next most acceptable alternative for all participants will have to be discussed. Facilitators should be wary of situations where consensus is reached very quickly and easily, as it may cover up differences of opinion which were not made explicit. It is important that any underlying 'political' issues are brought out into the open and dealt with honestly and explicitly.

3.3 WORKSHOP REPORTS

The facilitator is expected to present a report to the client on the outcome of each workshop. This report should be concise, summarising the main points and conclusions from the workshop and outlining the recommendations for implementation.

REPORT CONTENTS

> Workshop reports should include the following :
>
> - *After a workshop at concept stage (VM1):*
>   - a clear statement of project objectives, if possible structured and ranked in the form of a value hierarchy
>   - an initial set of design options which will be further evaluated and developed.
>   The report may, however, also question the need for a construction project.
>
> - *After a workshop at feasibility stage (VM2):*
>   - a clear recommendation on the design option offering the best value for money. This represents the point at which the outline brief becomes frozen; any further changes will lead to significant cost penalties, delays and disruption
>   - a weighted list of project objectives
>   - a list of suggestions for further improvement to the chosen design option.
>
> - *After workshops at the design stage:*
>   - a range of outline recommendations for design modifications and/or further design development. The client should carefully consider the changes recommended, as some of them may require considerable additional expenditure or may lead to disruption and delay.

There is much debate and disagreement on the ideal length and format of workshop reports. Generally, reports should be short and concentrate on the main issues. They should ideally follow the workshop sequence, have an executive summary, be well illustrated and contain necessary supporting detail in appendices.

TOOLBOX 4 EVALUATION TECHNIQUES

The following descriptions of structuring of objectives, the development of a value hierarchy, criteria/importance weighting and matrix analysis are all described in a paper by Stuart Green (one of the authors of this guide) as part of a SMART[1] (Simple Multi-Attribute Rating Technique) approach to Value Management.

[1] **Green, S.D (1992)** *A SMART methodology for value management.* See *Toolbox 6* for further details.

These techniques are aimed at establishing a common understanding of decision objectives and identifying possible solutions by way of a group consensus process. They provide an approach to decision making which is progressively revised in the light of discussion and debate.

4.1 STRUCTURING OBJECTIVES/DEVELOPING A VALUE HIERARCHY

The process of structuring project objectives and constructing a value hierarchy should be undertaken both during the workshop at concept stage (VM1) and the workshop at feasibility stage (VM2). It aims to **establish a shared perception of the design objectives and attributes.**

VALUE HIERARCHY AT VM1

> **AT VM1** the **objectives** identified during the first, information stage of the workshop (where participants present their interpretation of project objectives) are **structured into a value hierarchy** (see example in Figure A overleaf).
>
> The **procedure** is as follows:
>
> - the top of the hierarchy or 'tree' (left of figure) embodies the primary objective or 'raison d'être' of the project
> - this is progressively broken down into **sub-objectives** by way of a 'means-ends' analysis, i.e. the lower order objective, 'accessibility to customers' in the example overleaf, is a means to the primary objective, 'a successful retail development'
> - lower order objectives can then be further broken down into the next level of sub-objectives and so on. In the example, 'provision for car parking' is one of the means to the higher level objective of 'accessibility to customers'. Judgement is required to determine the extent to which objectives should be subdivided, but for most projects two levels of breakdown will be sufficient
> - the development of the value tree is an iterative process based on discussion and negotiation. Only when **group consensus** is reached and all participants accept the hierarchy as a fair representation of design objectives, is the process complete.
>
> It is important that **project objectives and constraints** are **kept separate** from each other. The value hierarchy should not include any constraints; these should be listed separately.
>
> Following the structuring of objectives, subsequent stages of the workshop at VM1 question how else the objectives can be achieved (this is done in a brainstorming format) and the ideas produced are then evaluated in terms of cost and feasibility. These stages are described in 4.3.2.

FIGURE A
**VALUE HIERARCHY
AT VM1 FOR
NEW RETAIL
DEVELOPMENT**

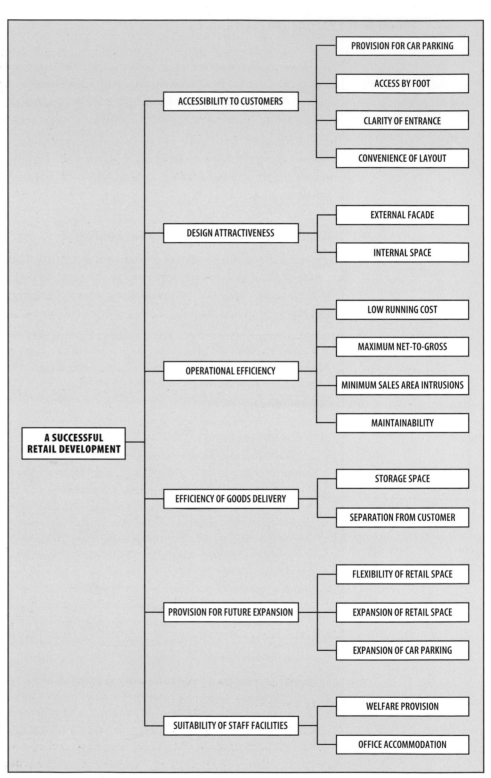

VALUE HIERARCHY
AT VM2

AT VM2 the *value hierarchy* drawn up during the first workshop is *restructured* in accordance with revised project objectives. These are identified during the information stage of the VM2 workshop, where participants comment on the extent to which project objectives and constraints identified at VM1 remain valid following changes that have taken place since then.

This time the *lower order objectives* are to be *used as attributes for the evaluation of alternative design options* during the next stage of the workshop, when 'weights' for the different attributes can be assigned (see *Toolbox 4.2*).

The restructuring of the value hierarchy involves the following *tasks:*

- the value hierarchy is *simplified* and attributes which do not directly influence the choice of design are eliminated (an example is given in Figure B below). The workshop should seek to answer the question 'will this attribute influence our choice of outline design?'. If the answer is 'no', the attribute should be taken off the hierarchy

- the facilitator should ensure that different attributes do not measure the same criterion

- attributes which are regarded as fundamental for *all* design options (e.g. attributes relating to safety) should also be 'pruned' from the hierarchy

Although *capital cost* will be one of the attributes used to evaluate design options, it is preferable to *omit* it *from the value hierarchy* and deal with it separately at the end of the analysis.

FIGURE B
**SIMPLIFIED VALUE
HIERARCHY AT VM2**

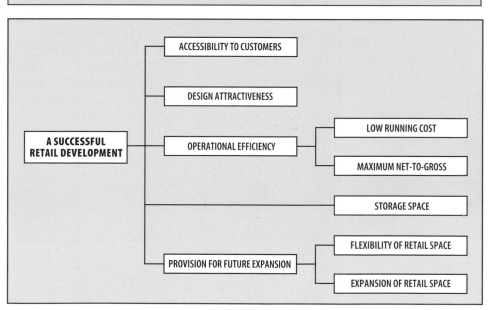

4.2 CRITERIA/IMPORTANCE WEIGHTING AND MATRIX ANALYSIS

4.2.1 CRITERIA/IMPORTANCE WEIGHTING

The process of criteria/importance weighting takes place during the workshop at feasibility stage (VM2). **Each attribute** of the simplified value hierarchy drawn up during the previous stage of the workshop (structuring of objectives) is **weighted according to its perceived importance** (see example in Figure C overleaf).

PROCEDURE FOR THE WEIGHTING OF ATTRIBUTES

Referring to Figure C below, attributes are 'weighted' as follows:

- first of all, the lower order attributes of the *first level* or 'branch' of the value tree are listed in order of their perceived importance and a *weight* of 10 awarded to the *least important attribute*. Weights are then allocated to the *other attributes* on the basis of their perceived *relative importance* in relation to the least important attribute. In the example overleaf, 'provision for future expansion' is seen as the least important attribute, receiving an initial weight of 10, whereas 'accessibility to customers' and 'design attractiveness' are seen as most important, each receiving initial weights of 50. The weights are then *summed* and for each attribute the *ratio* to this sum is *calculated and normalised* so that the total for the first level or 'branch' adds up to 1. In our example, the ratio of the attribute 'accessibility to customers' (weight of 50) to the sum of weights for that 'branch' (170) is calculated and normalised to produce an importance weighting of 0·29 (50:170)

- the *same procedure* is then applied to the lower order attributes at the second level or 'branch' of the value tree, with the total for each group (or 'branch') again adding up to 1. In the example below, at the *second level* 'net to gross' is seen as the least important attribute, with a weight of 10, and 'low running cost' as the most important, with a comparative weight of 40. Thus, when normalised, 'net to gross' receives a weighting of 0·20 (10:50) and 'low running cost' receives a weighting of 0·80 (40:50)

- finally, the importance weights are *multiplied* horizontally *through the tree* to provide the *final weighting* for each individual attribute. In our example, the weighting for 'operational efficiency' (0·24) is multiplied by the weighting for 'low running cost' (0·80) to provide a final weighting for the lower order attribute of 0·19. This is the weighting which will be applied to the decision matrix during the next stage of the workshop (see *Toolbox 4.2.2*)

- importance weights should be assigned on the basis of discussion and negotiation, i.e. on the basis of *group consensus* (*not* on the basis of voting) and the process is not complete until the weights are fully consistent and all participants agree on the final result.

FIGURE C
SIMPLIFIED VALUE TREE AT VM2 WITH IMPORTANCE WEIGHTS

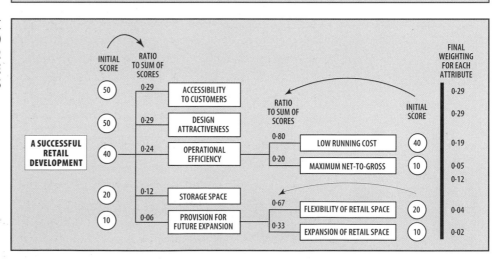

4.2.2 MATRIX ANALYSIS

The allocation of importance weights to the value hierarchy forms the basis for the next stage of the workshop at VM2, which is deciding which of the available options provides the best value. Evaluation involves *assessing each option against each of the identified attributes* and this is best done in the form of a *decision matrix* (see example in Figure D opposite).

DECISION MATRIX PROCEDURE

Referring to Figure D below, options can be assessed against attributes as follows:

- *each design option is scored against each attribute* on a scale from 0 to 100 (100 as the highest and 0 the lowest). In the example below, design option B scores best for 'accessibility to customers' (score of 90), followed by design option A (70) and design option C (50). Scores will frequently be allocated on a subjective basis, but where possible objective measures should also be used (these will have to be converted to a 0–100 scale)

- following the allocation of scores, these are then *weighted* (i.e. multiplied) *by the appropriate importance weighting* identified during the previous stage (assignment of importance weights) for each lower order attribute. In our example, design option A scores a total of 20 (shown in the shaded triangle) for the attribute 'accessibility to customers', i.e. score of 70 multiplied by weighting of 0·29

- the *weighted scores* for each attribute in design option A (i.e. 20, 5·8, 1·9, etc.) can then be *added together* to provide the aggregate score for design option A (36·5), which provides an indication of the overall value of design option A. *It is by the process of comparing the total scores of the various design options that the most suitable option can be arrived at* (e.g. design option B in Figure D).

FIGURE D
DECISION MATRIX FOR NEW RETAIL DEVELOPMENT

ASSESSMENT ATTRIBUTES	ACCESSIBILITY TO CUSTOMERS	DESIGN ATTRACTIVENESS	LOW RUNNING COST	MAXIMUM NET-TO-GROSS	STORAGE SPACE	FLEXIBILITY OF RETAIL SPACE	EXPANSION OF RETAIL SPACE	TOTAL
WEIGHT OF IMPORTANCE (0–1)	0·29	0·29	0·19	0·05	0·12	0·04	0·02	**TOTAL**
DESIGN OPTION A	70 20	20 5·8	10 1·9	20 1·0	40 4·8	60 2·4	30 0·6	36·5
DESIGN OPTION B	90 26	80 23	20 3·8	60 3·0	70 8·4	60 2·4	80 1·6	68·2
DESIGN OPTION C	50 15	50 15	80 15	40 2·0	10 1·2	40 1·6	40 0·8	50·6
etc.								

The sample matrix shown above could be simplified by reducing the design option scoring scale to 0–5 (instead of 0–100), with each number representing the performance of individual design options against individual assessment attributes as follows:

0 = UNACCEPTABLE   **1** = POOR   **2** = FAIR   **3** = GOOD   **4** = VERY GOOD   **5** = EXCELLENT

However, this may be too simplistic for certain circumstances and the methodology should therefore be adapted as required.

Subsequent stages of the VM2 workshop can *test* how sensitive the outcome of the above rating process is to *marginal changes in key variables* (e.g. examining how sensitive the results would be to changes in the importance weights). The estimated *costs* of each option should also be considered, which, when compared with the aggregate scores, provide an indication of relative value for money. The chosen option is then scrutinised by the workshop team in a *brainstorming session*. During this session, concerns are raised, then possible solutions and improvements suggested. To conclude, the ideas raised are evaluated and a summary list of issues requiring further investigation is produced. It should be noted that this kind of analysis relies heavily on subjective data and will not give precise results. Care should be taken when selecting options on the basis of relatively small differences in scores between alternatives.

4.3 FUNCTION ANALYSIS

The following descriptions of function analysis and life-cycle costing apply to Value Engineering and its objective of building cost-effectively, i.e. achieving the necessary function at minimum cost in cases where the subject of the exercise is well defined.

Function analysis is a simple concept which is often needlessly over-complicated. By focusing on the **function of a design or component** the technique helps to identify **alternative ways** of providing it. To begin, functions of the design being studied are examined by asking **'what does it do?'**. Functions are classified as either **basic** or **secondary**. Basic functions embody the purpose of the design component, and secondary functions are needed to support the basic function.

Function analysis is not a very precise technique, but it can help to **bring clarity** to a problem and to **develop a shared understanding** of the reason why design components exist or are needed. It is also a useful method of developing different perspectives on design options.

Function analysis is normally used during the first (information) stage of Value Engineering workshops. Parts of the design offering the greatest potential for eliminating unnecessary cost are identified and their functions examined. Design alternatives are generated during a speculation stage (brainstorming session) and are then examined in the third (evaluation) stage in the light of the functions identified during the information stage. It should be remembered that a key objective of Value Engineering is to reduce unnecessary cost **while maintaining the required function**.

FUNCTIONAL ANALYSIS
SYSTEMS TECHNIQUE

One form of functional analysis is the 'Functional Analysis Systems Technique' (FAST), which uses a function diagram, sometimes referred to as a Technical FAST Diagram (the procedure is represented in Figure E opposite and an example is provided in Figure F). A technical FAST diagram is constructed as follows:

- the **scope of the analysis** is defined by means of **two vertical dashed lines** to the left and right of the diagram. Everything between these two scope lines is included in the analysis

- all functions are defined using two words: **a verb and a noun**

- the **higher-order function** (the overall desired result or output) is identified and placed on the left-hand side of the diagram (outside the scope line)

- the **basic function** (which embodies the purpose of the design or component being studied) is placed next to the primary function immediately inside the left-hand side scope line

- all essential **design criteria** should be placed above the basic function in dotted boxes

- the **secondary functions** (or sub-functions) are then arranged to form a 'critical path', moving from left to right, so that each subsequent function answers the question of **how** the preceding function can be achieved. The logic of the diagram can be checked by reading from right to left and asking the question of **why** a particular component/function exists. This helps to verify the structure and validity of the diagram

- **secondary functions** which do not occur in a time-sequence, but occur **at the same time** as the other secondary functions in the critical path should be placed below the corresponding function on the critical path

- **secondary functions** which occur **all the time** should be placed in dotted boxes above the critical path at the right of the diagram (inside the scope line).

FIGURE E **FUNCTION ANALYSIS SYSTEM TECHNIQUE (FAST) DIAGRAM PROCEDURES**

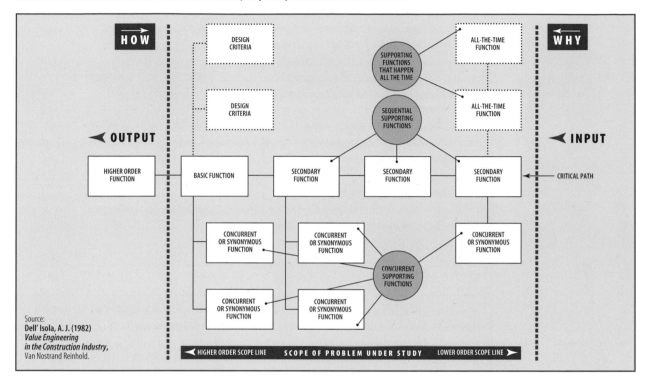

Source:
Dell' Isola, A. J. (1982)
*Value Engineering
in the Construction Industry,*
Van Nostrand Reinhold.

FIGURE F **FAST DIAGRAM - SHOPPING CENTRE**

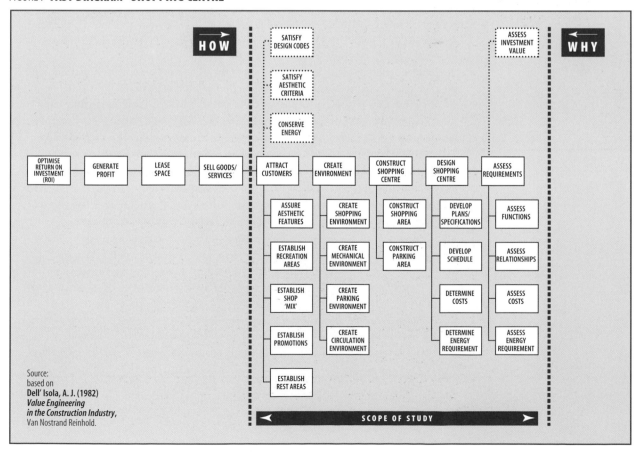

Source:
based on
Dell' Isola, A. J. (1982)
*Value Engineering
in the Construction Industry,*
Van Nostrand Reinhold.

There are a number of potential difficulties related to FAST diagrams:

- FAST diagrams were initially developed for use in manufacturing. The underlying logic is *not always transferable to construction projects*

- for many design components a large number of secondary functions can be identified and this can lead to an *over-complicated analysis and diagram*. It is here that the skill and judgement of the value manager/facilitator is very important in deciding what level of analysis is required

- much time can be wasted in attempting to solve a problem which has little impact on the higher-order function or which is caused by a technical solution which has limited potential for development. The value manager/facilitator should therefore ensure that the *discussions focus on the areas of greatest potential* and that they do *not lose sight of the higher-order function*

- the time involved in producing a FAST diagram may *alienate non-specialist participants*. Again the skill and judgement of the value manager/facilitator is required in deciding on the level of analysis necessary.

A second type of function diagram which is more readily applicable to construction is the *function hierarchy model*, also referred to as a *Task FAST diagram* (see example in Figure G opposite). While it follows the same 'how-why' logic of a Technical FAST diagram, it is not constrained by the same conventions and is easier to construct. For instance, the diagram does not have a time sequence.

The procedure to create a function hierarchy model is as follows:

- the higher-order function embodying the overall desired result or output is identified and placed on the left-hand side of the diagram (functions are defined in terms of a verb and a noun, as with Technical FAST diagrams)

- the basic functions embodying the purpose of the design or component being studied are then placed next to the higher-order function, immediately to the right of it. In our example there are several basic functions ('housing resources', 'facilitating study' etc.)

- these basic functions are then broken down into secondary functions (or sub-functions) which answer the question of *how* the basic function can be achieved. In our example the basic function of 'housing resources' can be achieved by 'storing text' or 'storing audio-visual material' while the basic function of 'facilitating study' can be achieved by 'accommodating students' and 'facilitating retrieval (of material)'

- secondary functions can then be broken down further into progressively lower levels of secondary functions to help explain *how* individual functions can be achieved.

- if the diagram is read from right to left it can answer the question of *why* a particular function is necessary - this helps to verify the structure and validity of the diagram. In the example opposite the question of why a library is required to have catalogue materials can be answered by the library's function to 'facilitate the retrieval' (of material) and in turn the function of 'facilitating study'

- the estimated costs of individual functions can be applied to the model, thereby providing a visual representation of the functional breakdown of expenditure. Those secondary functions which attract a high cost will constitute an obvious target for Value Engineering.

FIGURE G
**FUNCTION HIERARCHY
MODEL OF A LIBRARY**

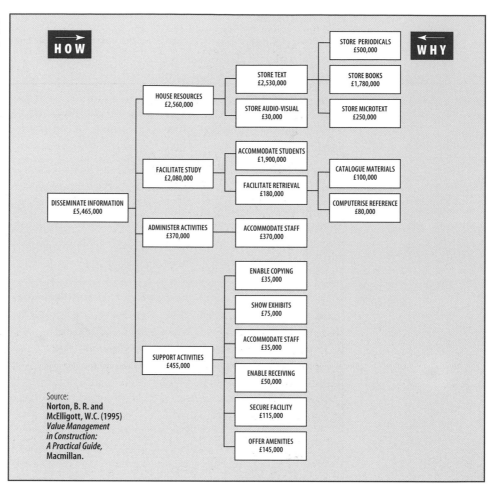

Source:
**Norton, B. R. and
McElligott, W.C. (1995)**
*Value Management
in Construction:
A Practical Guide,*
Macmillan.

A diagrammatic representation of a costed function hierarchy model can be very helpful when comparing the cost of functions of different options (see Figure H below). For cost comparisons to be meaningful, all functions have to be clearly defined. In practice the costing of functions can be quite difficult, *as costs may be influenced by external attributes* which are unrelated to the functions themselves.

FIGURE H   **COST COMPARISON OF ALTERNATIVE LIBRARIES**

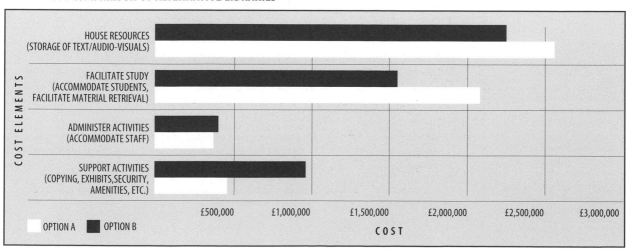

4.4   LIFE-CYCLE COSTING

Life-cycle costing is aimed at ***assessing the costs of a project or design over its lifetime.*** This takes account of initial capital costs, replacement/disposal costs, operating, maintenance and repair costs. Life-cycle costing can be used in Value Engineering to provide a meaningful comparison of the total cost of different design options. Anticipated future costs are converted to a common base by discounting future cash flows in order to identify the net present value (NPV) of each option. A simple life-cycle costing example is provided on page 70.

Life-cycle costing is usually applied during the development stage of a Value Engineering workshop. During this stage, the ideas carried forward from the previous stage (evaluation) are designed in more detail, examined for durability and buildability and costed. Note that to assess options on the basis of cost alone, each must have an equivalent performance. Note also that discounted cash flow relies on subjective judgements about future outcomes (interest rate movements, inflation, etc.).

LIFE-CYCLE COSTING PROCEDURE

> In a life-cycle costing exercise, the ***capital cost*** and ***likely future operating and maintenance costs*** of different design options are all ***estimated***. This involves the following steps:
>
> - first of all, all relevant cost ***elements*** are ***identified*** and well defined in a cost breakdown structure. An example of typical cost elements on a construction project is provided below
> - the ***costs*** of each element then are ***estimated*** from:
>   - known factors or rates
>   - the relationship with historical data (this type of estimating must be treated with caution)
>   - expert opinion (often the only method available when useful data cannot be obtained)
> - all the ***costs*** are then ***discounted***, i.e. adjusted to their present value by applying a discount rate to make the comparison of various alternatives possible and meaningful (see life-cycle costing example on page 70). This is necessary as different options will incur costs at various points in their life-cycle; discounting takes account of the value of money over time
> - ***inflation*** can be accounted for within the discount rate to be used (see inflation adjustment formula opposite). If the inflation rate is likely to differ substantially for major components, it may be necessary to discount the costs of these components separately.

The ***typical cost elements*** to be considered in a life-cycle costing exercise for a construction project are outlined below. The elements to be included will vary slightly from project to project.

CAPITAL COSTS	FINANCE COSTS	MAINTENANCE, REPLACEMENT AND ALTERATION COSTS	
– land acquisition	– finance for land purchase	– internal and external landscaping	– gas installation
– design team fees	– finance for construction works		– lift and conveyor installation
– demolition and site preparation	– finance during period of intended occupation	– external decorations	– communications installation
– construction works		– internal decorations	– special and protective installation
– statutory consents	**OPERATING COSTS**	– main structure and envelope	– external works
– development land tax	– local government charges	– finishes, fixtures and fittings	**RESIDUAL VALUES:**
– capital gains tax	– insurance	– plumbing and sanitary services	– resale value (building, land, plant, equipment)
– value added tax	– security and health	– heat source	– related costs (demolition and site clearance, disposal fees and charges)
– furnishing and fittings	– management and administration	– space heating and air treatment	– capital gains tax and balancing charges
– commissioning work	– energy	– electrical installation	
– decanting charges	– cleaning		

STANDARD LIFE-CYCLE COSTING FORMULAE
Before the life-cycle costing example can be explained it is necessary to outline the *standard formulae used in life-cycle costing exercises.* These are briefly outlined below. Further details are available in HM Treasury (1991) *Economic Appraisal in Central Government*, see *Toolbox 6*.

PRESENT VALUE OF £1 $= A \times \dfrac{1}{(1+i)^n}$

this formula shows the present value of a future accumulated amount, where:
A = *accumulated amount*
i = *interest rate (as a decimal)*
n = *time period (usually number of years)*

COMPOUND INTEREST $= P \times (1+i)^n$
this formula shows the future accumulated value of an initial capital, where:
P = *principal or present value (e.g. initial capital)*
i = *interest rate (as a decimal)*
n = *time period (usually number of years)*

SINKING FUND $= A \times \dfrac{i}{(1+i)^n - 1}$

this formula shows the amount of money to be put aside each year to cover a future known expenditure, where:
A = *accumulated amount*
i = *interest rate (as a decimal)*
n = *time period (usually number of years)*

LOAN REPAYMENT $= P \times \dfrac{(1+i)^n \times i}{(1+i)^n - 1}$

this formula shows the annual repayments due over a loan's life, where:
P = *principal or present value (i.e. size of loan)*
i = *interest rate (as a decimal)*
n = *time period (usually number of years)*

YEAR'S PURCHASE OR PRESENT VALUE OF £1 P.A. $= \dfrac{R\,[1 - (1+i)^{-n}]}{i}$

this formula shows the present value of an annual expenditure to take place over 'n' number of years, where:
R = *payments due at the end of each period (usually year)*
i = *interest rate (as a decimal)*
n = *time period (usually number of years)*

INFLATION ADJUSTMENT $= \dfrac{(1+t)}{(1+f)} - 1$

this formula shows the discount rate to be applied which takes account of inflation, where:
t = *actual discount rate (usually bank base rate)*
f = *inflation rate*
Note: to obtain a percentage the result from the formula has to be multiplied by 100

LIFE-CYCLE COSTING EXAMPLE **Question:** What are the comparative life-cycle costs of the following heating systems installations for a detached house?

BASIC DATA	SYSTEM A	SYSTEM B
Initial purchase cost	£ 1,500	£ 1,750
Annual maintenance cost	£ 50 p.a.	£ 50 p.a.
Pump replacement cost	(8th year) £ 400	(10th yr) £ 450
Annual fuel cost	£ 800 p.a.	£ 750 p.a.
Time period of the investment	15 years	15 years
Expected residual value of investment	£ 250	£ 250
Interest rate	7%	7%

Answer: All future expenditures have to be discounted, i.e. adjusted to their present value.

		SYSTEM A	SYSTEM B

1. INITIAL PURCHASE COST — £ 1,500 £ 1,750

2. ANNUAL MAINTENANCE COST:

It is known that for both systems the annual maintenance cost will be £ 50 p.a. over the time of the investment, i.e. 15 years, so calculate the total present value by applying the year's purchase formula:

$$\frac{R[1-(1+i)^{-n}]}{i} \quad \blacktriangleright \quad \frac{50[1-(1+0\cdot07)^{-15}]}{0\cdot07} \quad = \quad £\,455 \qquad £\,455$$

3. PUMP REPLACEMENT COST:

The pump will cost £ 400 to be replaced at the end of the 8th year for System A and £ 450 at the end of the 10th year for System B, so the present value of this future amount can be calculated by applying the present value of £1 formula:

$$A \times \frac{1}{(1+i)^n} \quad \blacktriangleright \quad \overset{\text{(SYSTEM A)}}{400 \times \frac{1}{(1+0\cdot07)^8}} \quad \overset{\text{(SYSTEM B)}}{450 \times \frac{1}{(1+0\cdot07)^{10}}} \quad = \quad £\,233 \qquad £\,229$$

4. ANNUAL FUEL COST:

It is known that the annual electricity cost will be £ 800 p.a. over the time of the investment for System A and £ 750 p.a. for System B, so calculate the total present value by applying the year's purchase formula:

$$\frac{R[1-(1+i)^{-n}]}{i} \quad \blacktriangleright \quad \overset{\text{(SYSTEM A)}}{\frac{800[1-(1+0\cdot07)^{-15}]}{0\cdot07}} \quad \overset{\text{(SYSTEM B)}}{\frac{750[1-(1+0\cdot07)^{-15}]}{0\cdot07}} \quad = \quad £\,7,286 \qquad £\,6,831$$

The total present values so far for the heating systems are — £ 9,474 £ 9,265

However, the residual value of the investment also has to be taken into account, i.e. its present value has to be subtracted from the above figure:

5. RESIDUAL VALUE:

The residual value of both heating installations at the end of the investment period (i.e. 15 years) is £250, so the present value of the future amount can be calculated by applying the present value of £1 formula:

$$A \times \frac{1}{(1+i)^n} \quad \blacktriangleright \quad 250 \times \frac{1}{(1+0\cdot07)^{15}} \quad = \quad £\,91 \qquad £\,91$$

TOTAL PRESENT VALUE OF INSTALLATIONS — £ 9,383 £ 9,174

Conclusion: System B would appear to be a better investment than System A over a 15 year period. This assumes that all costs and benefits are accounted for in the basic data. Changing this date – for example, the period of the investment or the interest rate – may affect the relative life-cycle costs of both systems.

TOOLBOX 5 TYPICAL COST ELEMENTS ON CONSTRUCTION PROJECTS

The table below shows a typical **breakdown of construction cost** for building and civil engineering works (an office building and a road project have been taken as examples). It would be inappropriate to derive averages for all works based on these breakdowns. Price ranges for the various elements are provided, indicating which elements typically incur a relatively high cost. The figures only provide an **approximate indication** of the likely cost.

BUILDING PROJECTS		CIVIL ENGINEERING PROJECTS (ROADS)	
Services	28–40%	Earthworks	28–31%
External envelope	15–18%	Structures	18–32%
Internal division	8–10%	Sub–base & surfacing	21–28%
Prelims, fees, site costs, etc.	6–9%	Drains & sewers	10–18%
Substructure	5–9%	Accommodation works	2–3%
Frame	4–7%	Kerbs & footways	1–3%
Roof	4–6%	Fences	1%
Floor finishes	3–4%	Traffic signs & road markings	1%
Ceiling finishes	2–3%	Lighting	1%
Upper floors	2–3%	Work for statutory bodies	0–1%
Wall finishes	2–3%		
Furniture & fittings	1–5%		
Stairs	1–2%		
Total building cost	**100%**	**Total cost**	**100%**

On building projects, **services** in particular account for a very large percentage of the overall cost (28–40%). This element can be further broken down into **mechanical services** (17–28%), **electrical services** (6–13%) and **lifts** (0–3%). On road projects the three highest cost elements (all roughly of the same order) are typically **earthworks** (28–31%), **structures** (18–32%) and **sub-base and surfacing** (21–28%).

The cost of external works for building projects (which is not included in the above breakdown of building costs) could typically account for an additional 7–18% of the total building cost.

TOOLBOX 6
FURTHER READING

This bibliography has been selected to provide a representative cross-section of useful literature. While the list is deliberately biased towards construction in the UK, the important American texts are also included. Brief reviews are provided for key selected texts.

GENERAL TEXTS ON VALUE ANALYSIS, VALUE MANAGEMENT, VALUE ENGINEERING:

Fowler, T.C. (1990), *Value Analysis in Design,* Van Nostrand Reinhold, New York.

Miles, L.D. (1972), *Techniques of Value Analysis and Engineering,* 2nd edition, McGraw-Hill, New York.

Parker, D.E. (1985), *Value Engineering Theory,* Lawrence Miles Foundation, Washington, D.C.

Snodgrass, T.J. and Kasi, M. (1986), *Function Analysis,* University of Wisconsin, USA.

VALUE MANAGEMENT AND VALUE ENGINEERING IN UK CONSTRUCTION:

Dallas, M.F. (1992), *'Value Management - Its Relevance to Managing Construction Projects.'* In: *Architectural Management* (ed. P. Nicholson), Spon, London, pp. 235–246.

Green, S.D. (1994), *'Beyond Value Engineering: SMART Value Management for Building Projects.'* In: *International Journal of Project Management,* Vol. 12, (1), pp. 49–56.

This paper argues that while assumptions implicit in 'traditional' Value Engineering are valid for well-defined technical problems, they are less so for the messy, dynamic and ill-defined social problems which characterise the early stages of building design. Value Engineering is equated with 'hard systems thinking', being primarily concerned with cost reduction. Value Management is equated with 'soft systems thinking', being concerned with developing a common understanding among project stakeholders. It is argued that the two approaches are applicable in different circumstances.

Green, S.D. (1992), *A SMART Methodology for Value Management, Occasional Paper No. 53,* Chartered Institute of Building, Ascot.

A simplified approach to multi-attribute utility theory, known as SMART, is recommended as an improved approach to Value Management during the early stages of building design. The approach is best suited to design situations which are characterised by multi-faceted clients and ambiguity. SMART is primarily perceived as an aid to the briefing process, rather than a cost reduction technique.

Green, S.D. and Popper, P.A. (1990), *Value Engineering: The Search for Unnecessary Cost, Occasional Paper No. 39,* Chartered Institute of Building, Ascot.

This paper provides a summary of the philosophy and methodology of traditional Value Engineering, noting that it is primarily concerned with the elimination of unnecessary cost. Three case studies of Value Engineering on UK construction projects are included.

Hayden, G. and Parsloe, C.J. (1996), *Value Engineering of Building Services, Publication AG 15/96,* Building Services Research and Information Association, Bracknell.

HM Treasury (1996), *Value Management, CUP Guidance Note No. 54,* Central Unit on Procurement, HM Treasury, London.

A short and practical guide which outlines how Value Management can be applied on public sector capital projects. It introduces the essential principles, and describes who should be involved and what may be achieved. It recognises that users will not be expert in the various techniques and identifies the need for expert help and advice.

Kelly, J.R. and Male, S.P. (1993), *Value Management in Design and Construction,* Spon, London.

Value Management is viewed as a service for maximising the 'functional value' of projects and the authors place great emphasis on functional analysis.
A number of case studies of North American Value Engineering practice are presented, as well as a useful introduction to group problem solving theory.

Kelly, J., Male, S. and MacPherson, S. (1993), *Value Management - A Proposed Practice Manual for the Briefing Process,* Research Paper No. 12, Royal Institution of Chartered Surveyors, London.

This paper proposes the use of Value Management studies for client briefing and includes three case studies which show how Value Management can be introduced at different project stages.

Locke, M.B. (1994), *'Value Management: A Blend of Process and Knowledge.'* In: *Proc. 5th European Value Management Conference,* Brighton, pp. 180–190.

Norton, B.R. and McElligott, W.C. (1995), *Value Management in Construction: A Practical Guide,* Macmillan, Basingstoke.

This book offers a comprehensive overview of Value Management. Value Management is subdivided into 'Value Planning' (applying to the early stages of design) and 'Value Engineering' (applying to the later stages). The book follows the traditional US model and succeeds in maintaining a practical flavour without ever being over-prescriptive.

Rwelamila, P. and Saville, P.W. (1994), *'Hybrid Value Engineering: The Challenge of Construction Project Management in the 1990s.'* In: *International Journal of Project Management,* Vol. 12, (3), pp. 157–164.

VALUE MANAGEMENT AND VALUE ENGINEERING IN US CONSTRUCTION:

Dell' Isola, A.J. (1982), *Value Engineering in the Construction Industry,* 3rd edition, Van Nostrand Reinhold, New York.

Ellegant, H. (1992), *'Modern Value Engineering for Design and Construction'.* In: *Architectural Management* (ed. P. Nicholson), Spon, London, pp. 247–255.

Ellegant draws the distinction between 'old Value Engineering' (primarily concerned with reducing cost as a 'tool of last resort') and 'modern Value Engineering' (primarily concerned with improving communication) which is perceived as an aid to the briefing process.

Macedo, M.C., Dobrow, P.V. and O'Rourke, J.J. (1978), *Value Management for Construction,* Wiley Interscience.

O'Brien, J.J. (1976), *Value Analysis in Design and Construction,* McGraw-Hill, New York.

Zimmerman, L.W. and Hart, G.D. (1982), *Value Engineering: A Practical Approach for Owners, Designers and Contractors,* Van Nostrand Reinhold, New York.

IMPLEMENTING VALUE MANAGEMENT IN ORGANISATIONS:

Isgar, P. and Drysdale, D. (1994), *'Introduction of Value Management into Railtrack.'* In: *Proc. 5th European Value Management Conference,* Brighton, pp. 134–146.

A practical paper which describes the current implementation of Value Management within the major projects division of Railtrack. The overall Value Management concept is subdivided into two areas: the application of Value Management techniques to 'soft' business issues (Value Management) and the application of value analysis to capital projects (Value Engineering). The paper describes the requirements for the successful introduction of a Value Management programme.

Quarterman, M.N. (1994), *'Implementing Value Management in BAA plc: A Case Study.'* In: *Proc. 5th European Value Management Conference,* Brighton, pp. 48–58.

The paper describes the BAA approach to Value Management, which is considered to be of prime importance during the early stages of project development. The role of Value Management is seen to be critical in challenging the need for projects and in establishing clear measurable objectives. In contrast, Value Engineering is seen to occur during the later stages of the project life-cycle and to be primarily concerned with the elimination of unnecessary cost. The 'critical success factors' for Value Management are seen to include the timing of the workshops, the quality of the initial information and the composition of the team.

Roberts, J.R. (1994), *'Value Management in the Process Industry.'* In: *Proc. 5th European Value Management Conference,* Brighton, pp. 70–101.

VALUE MANAGEMENT IN THE PUBLIC SECTOR:

Bone, C. (1992), *Achieving Value for Money in Local Government: Meeting the Charter's Challenge,* Longman, Harlow.

Bone considers value analysis as a tool capable of resolving a great variety of public sector management problems and views it as a line management function. He emphasises the importance of costing methods and performance measures.

Bone, C. (1993), *Value Management in the Public Sector - Through Value Analysis and BPR,* Longman, Harlow.

COMPANION CLIENT GUIDES FROM CIRIA

Connaughton, J.N. (1994), *Value by Competition – A Guide to the Competitive Procurement of Consultancy Services,* CIRIA, London.

Godfrey, P.S. (1995), *Control of Risk – A Guide to the Systematic Management of Risk from Construction,* CIRIA, London.

Potter, M. (1995), *Planning to Build? – A Practical Introduction to the Construction Process,* CIRIA, London.

Davis Langdon Consultancy and The University of Reading (forthcoming), *Value Management in UK Practice,* CIRIA, London.

MISCELLANEOUS

Chancellor of the Exchequer (1995), *Setting New Standards. A Strategy for Government Procurement,* HMSO, London.

Health & Safety Executive (1994), *Managing Construction for Health and Safety. Construction (Design and Management) Regulations 1994. Approved Code of Practice*, HSE Books, London.

HM Treasury (1991), *Economic Appraisal in Central Government. A Technical Guide for Government Departments,* HMSO, London.

Contents

Acknowledgments

HUGE thanks to the great girls who agreed to talk with me about growing up and what it means to them. It's not always easy to talk about periods and bodies. After all, that's pretty private stuff. But you all went above and beyond the call of duty to help other girls by explaining your feelings about some very personal things.

Preface

Girlchat

I am one amazingly lucky person! For years, as managing editor of *Teen Magazine*, (which had millions of teen readers) I got to be front-and-center when it came to the world of teens and preteens. It was the best job for someone like me who really likes people and has never lost touch with what it was like to be a teen.

From my cool, cozy little office on Sunset Boulevard (yeah—the one in Hollywood with the palm trees—does it get any better than that?) I was able to get up-close and personal with some very great girls all over the country (and even those from a buncha foreign countries, as well.) I read mailbags and mailbags of letters as well as e-mails from girls just like you. The letters were full of heartwarming stories and intelligent questions.

The phones rang nonstop at *Teen's* busy office, so sometimes I spoke with all kinds of teenage readers. I worked with glamorous teen models at photo shoots and met accomplished girls from all walks of life. I interviewed interesting teen celebrities who were known worldwide. What's more, I got to answer questions that girls like you asked when I wrote my advice columns, features, and health articles, as well as fashion and beauty articles.

When I left *Teen* to pursue other interests, I didn't stop talking to teens. Nu-uh. These days, I talk to them more than ever! And I still am in awe of you ladies. How could I help but be impressed by the teens of today? You girls absolutely, positively snap, crackle 'n' *rock*! You are compassionate, emotional, accomplished, aware of your world, and passionately interested in all kinds of important stuff. You're curious, and you ask all sorts of questions. You're not content to sit back and wait for info to come to you. No, no, nooooo. You get out there and seek out answers to all your most pressing questions!

And you're not shy either. You want to know more about what makes your body go, go, go. You want to know why you're growing and changing. You want to know about what it means to get your period and lots, lots more.

That's where I struck it lucky again and got to write this book. Once again, I had the chance to speak with real girls just like you who were only too happy to talk about the stuff you want to know about. These girls know what it's like to deal with changing bodies and feeling somewhat surprised, like they're . . . well, *on the spot* at times!

So make yourself comfortable. Grab this book and get ready for some bodywise girlchat that will help you as you go about being the best you yet!

Introduction

Y ou already know it's great to be a girl. Chances are, you're already well on your way to figuring out just how special you are. No matter what your personality—whether you're artsy, sporty, brainy, punk, a dreamer, a doer, or a combo of all of these—you've seen and experienced some very cool things in your life.

Maybe . . .

- ✿ You've got some very special friends who know just how to make you smile. Whether you hang out in a big group or gather a few close gal pals around you, you're never happier than when you're with your best buds.
- ✿ You're super close to your family, and you feel so loved, you just want to shout it out to the world. You know that even if you go through tough times, your family is always there for you.
- ✿ You've got a hobby that you like to do so much you can't imagine wanting to do anything else. Whether it's collecting stuffed animals or writing poetry, you're tickled that you've found something that interests you. You want to learn more about it.
- ✿ You've been lucky enough to travel to some way cool places like Disney World, the Grand Canyon, or maybe even Europe. You've learned a little about different places and have seen some fascinating sights.
- ✿ You've had opportunities to lend a helping hand to others and have found that there's nothing that makes you feel sooo alive!

Whether you've collected canned goods to feed hungry people or helped tutor needy kids, you know there's nothing quite like knowing you've made someone else's life a little easier.

But guess what? All these good things that you're experiencing are just the beginning! There are also some exciting physical changes ahead of you. Okay, okay, so some might be more exciting than others. Sometimes these physical changes don't seem so exciting—more like *excruciating!* But they are an important part of what makes you a girl—of what makes you uniquely, totally you!

You probably already have a clue that you won't stay in your little girl body forever. After all, you have eyeballs. You can't help noticing that the women around you look, well, like *women*. They've undergone some definite changes—and so will you.

One of these changes includes menstruating, a.k.a. getting your period. Like most girls, you're probably already looking ahead to this event with a mixture of anticipation and, well, maybe a not-so-slight case of n-n-n-nerves.

News flash: You don't have to go on worry overload about what's to come. *On the Spot* was written for girls like you who want to pump up their girlpower by being informed! It contains all you need to know about your period—and what it means to you.

What's more, this book is more than just a buncha super-techie scientific stuff—it probably goes beyond what you may have learned in health class. While scientific explanations are definitely important, let's face it, there's just more to growing up than sometimes boring body talk. *On the Spot* dishes up important facts, but it doesn't stop there. It's also jammed full of real-life experiences and tips from girls who are going through this amazing journey to womanhood—just like you are. These girls have been through some of the changes you'll be experiencing, and they weren't shy about speaking up. They were glad to help girls like you avoid some of the period pitfalls they encountered and to help ease your freaked-out fears.

By the time you finish *On the Spot*, you'll have gotten most of the physical and emotional growing-up goods you need to know. Why is it super important to be in the know when it comes to personal things such as periods? Here are just a few reasons:

✿ Girls who are armed with information don't feel like everyone knows something they don't know. They don't have to wonder what they're missing.

My older sister has some strange things she keeps in the bathroom. But whenever I ask her about them, she tells me to mind my own business. I want to know what's going on. — Jordan

Even though Jordan's body hasn't started changing yet, she's noticing that her big sister is changing in ways Jordan doesn't quite understand. It's easy to see that Jordan's ready for more information, and she feels a little left out. If you're like a lot of girls, you probably don't like feeling that everyone is in on a secret—except you.

✿ Girls who are familiar with what to expect don't worry about unknowns as much as other girls. They don't have to feel fearful when something they've never experienced before comes up. They already have enough information to feel comfortable with new situations. Even if they have questions about a detail, they have enough knowledge to keep most of their worries at bay. This gives a girl more confidence.

✿ Girls who are informed don't feel the need to keep quiet when something big happens. They don't feel alone.

✿ Girls who take time to learn about their bodies begin to find that even though they feel weird about asking questions at first, eventually they become comfortable enough to ask more questions. As a result, they learn more. This enables them to feel more sure of themselves and take more responsibility for their health and well-being. Can you say major c-o-o-l?

As you're reading *On the Spot*, or when you're finished, use this book as a springboard for talking with a trusted adult. Trusted adults can include your mom, a caregiver, an aunt that you're close to, or a family friend. Other trusted adults include health-care professionals, such as a family doctor, pediatrician, gynecologist, or nurse practitioner. A gynecologist is a doctor who practices the branch of medicine related to women's health.

When I was younger, I didn't really want to ask about stuff like menstruating. When my mom approached me and started talking, I admit I felt really strange at first. But then it got easier. — Ali

Like Ali, you'll find that once you get started, it's not so hard to talk about your health concerns and what growing up means to you. So, what are you waiting for? Gear up and start reading all about the Next Big Thing in your life!

Getting Bodywise

Getting bodywise is being open and honest about your health and what growing up means to you. This can be easy for some girls and tough for others. When it comes to learning to be bodywise, what kind of girl are you?

Eager Emily

You want to learn whatever you can about your body and what makes it special. You tend to listen carefully in health class. You're one of the girls who actually scans the brochures your teachers hand out. Scientific explanations don't scare you. In fact, you find them kinda interesting. You're generally the first one to ask questions when something's on your mind. You take mental notes when your family doctor talks to you about your health concerns.

Ignore-It Imogene

You're the kind to turn away when a trusted adult discusses health-related things. You tune out in science class if the teacher starts with the body talk. Most of the time your eyes glaze over in boredom when your friends ask questions about what's going on with their bodies. You figure if nothing's happening to you yet, there's absolutely no need to think about it.

Fearful Frieda

You start furiously doodling in your notebook in embarrassment if someone starts talking about health issues. *Ewwwwww! Who wants to even consider such stuff?* you think. You find yourself closing your ears and your mind as your family doctor brings up the subject of growing up. Any time you even *think* about how you might change one day, you get the willies. When it gets down to it, you just wish you could close your eyes and make the growing-up monster go away!

Okay, so not everyone is going to become an Eager Emily overnight. It just might take a little time to get used to the idea that things aren't always going to stay the same.

Still not convinced? Try thinking of your body as a sleek, high-performance machine, like a car or a computer. You're already aware that your body can do some great things. It's enabled you to experience the world in so many ways. If you're in the driver's seat, you'll be better able to take care of your body. Consider this book an owner's manual. Now kick your attitude into "go mode" and plan on learning all you can about yourself!

Meet the Girls!

Throughout this book, you'll meet some way-cool girls from all over the country who will share their experiences, thoughts, and advice on the subject of growing up. Ranging in age from nine to seventeen, these super chicks will offer you the Girls'-Eye View—the real scoop—on the things you need to know about your changing body. While their names have been changed for privacy, their thoughts, wisdom, and knowledge are straight from the heart.

Jordan, 9

Jordan is the original family girl, and she looks up to her parents and older sister. Around the house, she likes nothing more than to sing and dance. She may be young, but she's already thinking ahead. She's happy-go-lucky and always ready to explore her world. While right now she's just beginning to ask questions, there's no doubt that one day soon, she'll have the answers she's looking for.

Kacie, 14

Kacie is a thoughtful girl on the go. She's active in school and tries hard to make the grade, that's for sure. Though she likes English and history, she finds she does better in math and science. When school's out, though, she puts the thought of books out of her mind, and she does what she loves most—riding horses!

Crissy, 13

Crissy is musically inclined and loves to be where things are happening. Like Kacie, Crissy likes to ride horses and takes jumping lessons as often as she can. But she has other interests as well. She practices playing her clarinet and likes marching in the school band. She has a close circle of friends, many of whom are in band with her, and she's fiercely loyal. She also believes in helping others.

Libby, 10

With Libby, friends come first. Forget just calling 'em—Libby likes to fill her house with her amigas whenever she can. That's because she wants to be face-to-face with her best buds. When it comes to other interests, Libby likes active sports such as soccer and tennis. A voracious reader, she also collects glass animals and has a huge display of the crystal critters in her room.

Jenny, 16

She's the kind of nonstop girl who has big dreams of traveling all over the world. She's made a big list of all the places she'd like to visit, including Egypt to see the pyramids and Australia to see kangaroos. She works hard at school and participates in sports. Her real love, however, is music. She likes to listen to music, and she participates in several a cappella singing groups. (These groups sing without instrumental accompaniment.) She's also artistic and loves to paint. She tends to seek out upbeat friends who are adventurous and who like to have a good time.

Rachel, 17

Once you get to know Rachel, you know she's a true-blue buddy. Talk about caring—Rachel is the kind who'll stay up late at night to talk to a friend in need. She'll be the first one to arrive at your house with a little cheer-up present and a listening ear if she thinks you're feeling blue. Right now, she's busy working on her school's annual. She's in charge of gathering pictures and figuring out how to place them just right. She is steady and thoughtful, and uses her experiences as learning tools.

Alex, 14

Alex can't seem to sit still. Most mornings, she's off and running bright and early with her cross-country team. Sometimes she travels all over the country to lead the pack at track meets. Other times, you can find her showing her beloved champion dog, Tango, at dog shows. Though Tango's won quite a few ribbons, Alex prizes his "snuggability" most of all! When Alex winds down at the end of the day, you might find her at her easel creating pieces of art or at her computer screen hard at work designing Web pages.

Becca, 13

Becca for President? It's quite possible, so watch for this go-get-'em girl's name on a ballot one day. She's already won several coveted electoral offices at her school and now sits on the student council. She's learning a lot about leadership and how to put together great student activities. When she's not busy working on committees, you can find her on the sport courts. But when it's time to chill, Becca she loves nothing more than hanging out with friends.

Ava, 17

If you had to describe Ava in one word, "independent" would probably be it. Ava knows just where she's going, and how she's going to get there. Definitely outspoken, she's interested in so many things, she's not sure just what to do next. People are drawn to her endless energy, and they admire her generous spirit.

Trish, 14

No doubt about it, Trish has star power! She loves the performing arts—especially singing—and sees herself on stage one day. She likes participating in school productions and is interested in learning all she can about theater. She enjoys cranking up her music and singing along. She claims that while she's cool with the idea of school, she especially looks forward to lunchtime because that's when she can be with her friends.

Katherine, 16

Parlez-vous Français? Katherine does, and she wants to put her knowledge of French to use one day as an international businesswoman. For now, though, she's a high-school girl who loves staying close to home and being with her family. When Katherine's out and about, you can usually spot her carrying a camera. Why? Because she is very interested in photography. What's more, she enjoys volunteering and especially treasures the time she spends helping disabled children.

Ali, 15

Ali loves to make a splash. That's because one of her favorite sports is swimming. She likes the freedom that the water offers and the good feeling she gets from exercising. After she climbs out of the pool, she likes being with teammates and friends. She's a movie buff who often can be found with her buds at her community's local multiplex.

Emma, 13

As a singer, this girl hits all the right notes. She's a member of a select community choral group and rehearses for hours. She and her group are planning to head for Europe, where they'll sing for audiences in several countries. Emma's not one to limit herself, either. She's eager to learn all she can about a number of subjects and makes it a point to study and learn. Emma likes to soak in the scenery around her, so you can often find her taking long walks and appreciating Mother Nature.

Jemma, 13

This girl is packed with personality. She's talkative and inquisitive and has a sense of humor that just won't quit. Never one to sit still, she involves herself with lots of activities including sports. She enjoys working on her computer, but she also enjoys using it to socialize. That's because her favorite thing to do is to be with her friends, and her computer helps her keep in touch when she can't be with them in person!

1

Growing Up: Why Me?

Okay, so it's a little scary, but lots of girls like the idea that they're going to change one day soon. Maybe you're a kick-it-into-gear girl who can't wait to see what's next. Sure, you've enjoyed your childhood, but you're kind of psyched by the idea of becoming older and taking on some womanly ways.

If you're like Libby, who's ten, or Jordan, who's nine, you might look forward to having some of the goodies associated with being grown up.

"I think it'll be nice to grow older because people treat you with more respect," says Libby.

"Sometimes kids get brushed aside, but when you're older that won't happen so much anymore. People take you a little more seriously," says Jordan, who looks at growing up as an opportunity for freedom. "When you're older, you get more privileges," she says. "You don't have to go everywhere with your parents. You can go to more places you like, like the mall."

Okay, so the mall might not be exactly your definition of freedom. You might define it in different ways. But you probably can appreciate that entering teen territory means gaining more privileges. Who can help liking the idea of being treated more like an adult than a kid?

You might be looking forward to having a more womanly body as well. Jordan sure does:

Growing up is going to be so cool. I think it'll be nice to have some curves instead of being straight up-and-down.

Or, like a lot of girls, you might be more than a little freaked out and uncomfortable about the changes your body is about to go through. Katherine, who saw a video on growing up and getting your period when she was ten, recalls how she felt:

"I was shocked at first," she recalls. "I thought, Omigosh, this is going to happen to me? My friends and I started giggling. It wasn't that it was funny; it was to cover up how weirded out we were feeling."

But comfortable with the subject or not, a big part of growing up is realizing that change is coming.

All Grown Up?

Maybe you feel that growing and changing means that now you're mature. As you look around you at some of the girls whose bodies definitely are in the curvier category, you see that they certainly act like they think they're pretty grown up. And perhaps you think that the day your period arrives marks the day that you'll officially be a woman. Actually, that's kinda true, but not entirely. Your body will change, and your emotions will go through some growth as well. People might even treat you with more respect because you look more like a woman.

Girl's-Eye View: How You Mature

Keep in mind that maturing and being mature are two different things. Your bod might look womanly, but your emotions will still need time to play the catch-up game.

Jemma was eleven when she noticed the girls in her grade starting to act as if they were mature beyond their years. "They were dressing older and acting all girly," she says. "But the rest of us thought it was weird because it didn't make them any more mature than they really were."

If you're not quite sure this is true, give those curvy girls at your school a second glance. Some of them might be giving off some attitude now that their bodies are taking shape. They might say things that seem grown up, but really don't make much sense when you think about it. They've got some *real* growing up to do, even though right now they might not see it.

Luckily, lots of girls realize what growing up means and anticipate the day when they can confidently take more charge of their lives. And they look at physical changes merely as messages that they're on their way!

Girl's-Eye View: More Than Meets the Eye
When you get right down to it, growing up is more than having breasts and shapely hips. It's definitely more than physically being able to make a baby. It's about taking more responsibility for yourself, your health, and your actions.

Of course, when it comes to looking forward to change, you might be quite the opposite. Maybe you think that going through all the hassle of having your body morph almost magically, practically overnight, is just not for you. You're in no hurry, you say. You don't really see the point, and anyway, you like things just the way they are right now, thank you very much!

Chances are, however, that you've already gotten the sneaking suspicion that you don't have much of a choice in the matter. You kinda get the feeling that ready or not, your body is going to take you on this wild ride. The thing is, you can take your time in your own mind to get used to the idea, so that when changes start, you will be ready.

When things are happening so fast that you feel like you're trying to take a drink from an open fire hydrant, take a deep breath and relax. Keep your sense of humor, keep an open mind, and make the most of this once-in-a-lifetime experience.

Even though you'll be armed with lots of awesome information once you're finished with this book, don't be surprised if you still have mixed feelings about the matter of growing up. If you feel glad you're growing up yet freaked at the same time, don't worry. That's all part of the deal. It's perfectly normal.

Which brings us to the next question: Just exactly what is "normal," anyway?

Bra Banter

While on one hand you might look forward to wearing a bra, you don't want to go through the hassle of telling your mom you need one. Maybe you're already picturing her taking you to the store and loudly announcing to the saleslady, *"My daughter needs a bra!"* The thought makes you want to vaporize!

Shape Shifters

Could be you kinda like the idea of having a shapely body one day. But you're not sure if this means you'll like it if the boys around you notice that you've changed. And at the same time, you worry that maybe the boys will never notice you at all!

Period Pressure

Maybe you think it might be cool to have your period for the first time so you could get the experience and anxiety over with. Yet—yikes!—you don't want to deal with being prepared for periods for the next thirty-five or so years!

"I think it's important to know that even though your body is changing, this change is totally normal. Everyone goes through it. So you shouldn't worry about it. If you find yourself worrying just a little about something, you should remind yourself that you're perfectly okay." — Crissy

2

Your World: What's Normal?

If you're like a whole lotta girls, you probably think a *bunch* about what's normal and what isn't. This is especially true when you enter your teens. It's around this stage of your life that you step up your quest to get to know the world and figure out just where you fit in.

When you were a little girl, you were experiencing some growing-up things, that's for sure. You grew gradually from babyhood to toddlerhood and then on to childhood. Your baby face changed into a more childlike face. You lost your baby fat and became a little straighter in build. You lost baby teeth and got new, bigger teeth. All of these were signs that you were growing.

But the thing is, these changes happened so slowly, you didn't really pay much attention to them. They were just part of the background. You went about your life full speed ahead without looking back at how you changed along the way. The difference is that when you venture into teen territory, you start becoming hyperaware of the new things you're going through.

"The second I started sixth grade, I could tell things had changed," Ali, who's fifteen, says. "Whereas before the girls wore just regular T-shirts, now they were wearing all kinds of things that were trendy. A few girls got leather jackets. The rest of us couldn't see what was so wrong with wearing what we always had."

It's often at this time that you start considering outward appearances more than ever. Like Jemma, maybe you've noticed out of the corner of your eye that the girls around you are putting on a little more of a show than ever before. Before, your pals didn't care if their hair was perfectly parted. It probably didn't bother them if it was parted at all. They didn't mind tossing their hair back in a quick ponytail if they were in a hurry. They tossed on whatever clothes were comfortable and that was that. Looks just weren't that big a deal.

But these days it's as if someone waved a magic wand. Your friends seem to think of nothing but how they appear to the outside world. No matter how much you plead with your pals to step on it, they spend a half hour in front of their locker mirrors combing their long locks! They fret about their features and try to figure out how to improve their looks. Could be some of them are even starting to wear lip gloss and painting their nails.

When I was about twelve, I noticed that some of the girls at my school began thinking more about what they wore and wearing clothes with more style," says Trish, who's now fourteen. "It was the first time that I saw possibilities for myself. I liked the idea that there were things I could do to look better.

Like Trish, you might notice that lots of your friends are more finicky about fashion than before. This might start you thinking about your clothing choices as well. You might start feeling like a fashion failure because you're still wearing the same styles of clothing you've worn since you were in the primary grades. You might start begging your mom to let you wear clothes like your friends wear.

"At the beginning of middle school, I started seeing that my friends were becoming obsessed with their appearance," Becca recalls. "Every time I'd turn around, it seemed like they were ducking into the restroom to check up on their looks."

At the same time, you might find that you're focusing a little more on the glitzed-up girls on television, in music videos, at the movies, or in magazines. It's kind of hard to get away from these images.

They smile at you on TV screens at the store. They leap out at you from billboards as you're driving to gymnastics or karate practice. Sometimes you even find yourself staring at the latest fabulous face on your cereal box as you scarf down breakfast in the morning!

And all of this exposure kind of plays upon your mind. Maybe you think that because society splashes media princesses all over the silver screen and all over glossy magazine pages, these girls must be more worthy than you are somehow.

No one could blame you for thinking this way. Advertisers spend a great deal of energy to make you wish you looked different. They pay models lots of money to look a certain way. The girls go to great lengths to make sure their hair's perfect and their skin looks flawless. They're dressed to kill, and they look totally poised all the time. Even famous athletes seem to look perfect when they're smiling at you from the pages of a sports magazine!

Mirror, Mirror

Perhaps you find yourself thinking that you're supposed to look like a clone of the glam girls you see everywhere. Yet, when you look in the mirror, you happen to see someone

. . . with stick-straight hair that falls flat no matter what you do.

. . . or with wildly curly hair that sticks out all over the place.

. . . with teeth that are anything but even—or buried under braces.

. . . with bony knees that are covered with bandages.

. . . with skin that has its share of imperfections.

. . . with a straight body that resembles, well, a plank!

"Ooooookay," you might say to yourself. "I guess I'm anything but normal, after all. I look nothing like those glam girls do." Wrong answer!

Before you waste even one more minute putting yourself down, remember this: you—not they—are normal. You are a real, three-dimensional person. You're not a face plastered on a page or a video image captured on film for just a few minutes. You don't have an army of hair and makeup people working on you for hours like they do. (That's right—*hours*!) You don't have a camera crew messing around with angles and lighting to make sure you look flawless. You don't have computer people at work enhancing and changing your image. In short, you don't have people working twenty-four/seven to rub out your uniqueness!

And guess what? That's very good news! After all, when you get right down to it, you don't want to be like everyone else's idea of what you ought to be. Admit it, don't you like being just a little bit different, being just a little bit special?

C'mon. Have a little fun. Say it aloud: YES, I DO!

Jemma's Journey

Jemma, who's thirteen, shared this story of how she learned that being different was something to be celebrated.

"When I was in fifth grade, I felt a little unsure of myself. The girls around me were changing. Over the summer, they had really become much more interested in their appearance. I hadn't really thought about my looks that much before.

"I decided that I maybe I ought to tweak myself so I could fit in. I just didn't know how to do that. I looked around and saw that the girls in my class started dressing in certain ways. Okay, I decided, it was all about how you dressed. It was like there were certain categories that these girls all fit into. So I experimented by trying out the different styles I saw them wearing.

"Day after day, I'd show up at school in clothes that fit into each type. None of them seemed right for me. I just didn't feel like myself. My friends didn't respond well either to my new 'looks.' That's when I finally decided that I didn't fit into any categories. I finally figured out that it was just okay to be me!"

Rachel's Realization

Rachel, who's seventeen, tried modeling during her early teen years. And she found out that while modeling can be fun, it's way more important to maintain your individuality than to try to fit into someone else's idea of how you should be.

"I thought it would be exciting to model, so I got started," Rachel tells us. "At modeling school, I was shown how to stand straighter and not slouch. I was taught to apply makeup. My eyebrows were plucked (ouch!). Before I knew it, someone was cutting my hair. I was told to dress in a certain way, even though the clothes I was made to wear were nothing like I'd pick out on my own.

"I'd go on modeling calls (called *look-sees*) and then on photo-shoots. One person would tell me to wear my hair in yet a different way. Another person would completely redo my makeup. I'd be given different clothes to wear, and someone would style the clothes in ways I'd never dream of. It seemed like everyone had a vision of how I should look. I'd sit there in front of the camera, feeling like a robot, like my personality had been completely blotted out.

"After my sessions, I'd look at the pictures people had taken of me. I have to say, it was surreal. This strange face that I hardly recognized stared back at me. It was like I was a cookie-cutter person! I didn't look like myself at all. I looked like everyone else. In real life, I was me—different, quirky, maybe, but *me*! And I decided that I *liked* me, and that I'd no longer let someone else decide how I should look."

Looks, of course, are only a tiny part of who you are. What's going on behind that pretty face of yours is really what being you is all about. You need to look closer at who you are and what sets you apart.

So the first thing you need to do as you set about exploring the unique you and the changes you're undergoing is to remind yourself that the real you is cool. So what if the real you is a person who sometimes has less-than-perfect hair or knobby knees or a crooked smile? Nobody's perfect—not even those glam models.

Girl's-Eye View: Finding the Real You

The real you is a girl unlike any other girl in the world. She's a girl who's worthy of learning all she can so she can take great care of herself.

As mentioned before, this book talks a lot about the physical changes you're going to go through. But it also has a great deal to say about the inner you as well. That's because people are more than the sum of their body parts. You are a complex creature complete with a mind and emotions as well as a body. Your brain, your body, and your emotions all work together to make the unique you. And these aspects of you come into play even more when you hit puberty—that time in your life when things start to change.

"I believe you will find the whole growing-up thing will go a lot smoother if you decide to just take everything as it comes. Even if you feel awkward at times, try to remind yourself that everyone has these moments. Everyone has awkward moves here and there. Relax. Be a support to your friends, but don't make their worries your worries."   — Rachel

3

Your Changing Body: What's Up?

Changes Everywhere!

Okay, you say. You're convinced. You're glad you're a girl with your own special qualities, a girl who doesn't need to look like a movie star to feel good about herself. And though maybe these days you pay a little more attention to your appearance than you did before, you still strive to shine on your own. Still . . . you can't *help* but peek at your friends and make sure that overall, you fit in. You can't help but feel pretty psyched that you look more or less like the people around you.

But one day, just as you're starting to get a handle on being the girl you think you are, some things start to happen that jolt your world. It's called puberty, and it affects everything from how you look to how you feel. Maybe it starts in middle school. Like Katherine, you start noticing that some of the girls have bodies that are beginning to look different from yours:

At my school, it seemed like it happened overnight. One day we were all more or less the same. But then suddenly, the girls whose bodies were changing started sticking together. They seemed to split off from the rest of us. It was like this exclusive clique. If you weren't changing, forget it. You were just kind of left behind.

On the other hand, maybe you're the one who's hurtling ahead of your friends. While once you were growing up in a gradual sort of way, suddenly you notice that someone seems to have set your growth on fast-forward! Your body's the one that's going berserko! You may be changing in ways that seem to set you apart from your pals. Do any of these situations sound familiar?

Height Hopes

Say you've always been pretty average in height. When you line up for your class photo, you're usually somewhere comfortably in the middle. But one day, you're walking through the hallways at your elementary or middle school, and you start noticing that you're gaining on your friends. It's not so much at first; maybe you're slightly taller. But over the next few months, you notice that you continue to grow, and soon you're towering over most of your pals. *Whoa!* you think. You're becoming way taller than many of the boys at your school!

Growth Spurt

Maybe you're suddenly becoming aware that your arms are growing like crazy. Your sweater sleeves seem shorter than ever these days. And you wonder what's going on with your jeans. The other day, they fit just fine. But now the hems have climbed up your ankles, and your friends ask you if you're preparing for a flood. Your legs seem to stretch on forever. Why, you wail, can't you have normal arms and legs like everyone else you know?!!! You feel like you're all out of proportion and that you must look like a freak!

Shoe Blues

Sure, maybe you like shoe shopping as much as other girls. But these days, you're at the shoe store way more often than you want to be. That's because your feet seem to be getting bigger by the day. They just won't stop growing. In comparison with your body, your feet look way out of whack. At this rate, you wonder if you're going to end up resembling a clown—the kind with those oversized feet that look like boats!

Oh, No! B.O.!

You're out on the playing field, and you've just finished a high-energy game. You and your buddies are crowding into the locker room together and your nostrils fill with an unmistakable, unpleasant, um, odor! It smells like unwashed bodies, and you can't help but realize that you're contributing to the, well, *atmosphere* yourself. It's not the sweet smell of success—it's the unfriendly fragrance known as *eau de sweat*!

Bosom Woes

There you are, changing your clothes for gym class. As you glance down, you notice something that you hadn't seen before—you're getting breasts. Sneaking a look around you, you see that no one else's chest is changing. Your pals are as flat as DVDs while you're blossoming in a way that you're not sure you want to. Worse yet, you feel like people are looking at you.

Breast Distress

Or, the opposite breast thing is going on. You look up in the locker room and see that a bunch of your buddies are getting breasts—and bras. Not you. All you're getting is a big old case of envy. That's 'cause your breasts aren't growing at all! You're sure you'd prove the old joke to be true: If you got a living bra, it would die of starvation!

Girl Gab: Ahead of the Curve—or Behind It?

What if these changes start happening to you? What if you feel that you've gone from normal to, well, *weird*—and that you'll never be the same again? Should you freak if you're not like your friends? No way! Here's what real girls have to say about changing and growing—ahead of the curve, or behind it!

Tall Tale

Finding that you're zooming up in height? If so, stand tall and be proud. Rachel knows how you feel:

"Ever since elementary school, I have always been tall. I was even taller than most of the boys. I'd finally gotten used to it. Then my first year in middle school, I started growing like crazy. I just kept getting taller and taller. I was pretty self-conscious about towering over everyone. When it came to school dances, it bugged me that I was taller than the boys I danced with. I felt myself starting to slouch to try to make up for being such a giant."

Short Story

Feeling small because you've come out on the short end of the height stick? No matter. You've got BIG dreams, and that's what counts. Just ask Kacie.

“I was always the shortest kid in my class. When I starting shooting up in height in middle school, I was totally happy. *It's about time,* I thought. I had started to catch up with my friends! But then they started growing all of a sudden and getting pretty tall as well. It wasn't too long before they were taller than I was. Oh well. At least I wasn't as short as I had been before.”

Coordination Hibernation

It's not fun to have to take a coordination time-out, that's for sure. But be patient. You're sure to find your groove again soon.

"I had always been pretty coordinated. I loved to play sports and was used to being kind of good at whatever I played. But starting the summer when I was going into sixth grade, I started noticing that I was growing. My body was changing, and suddenly I didn't seem to know my way around anymore. I was this amazing klutz. I'd get my feet kind of tangled up when I was playing. My teammates would bag on me at times because I'd flub up a play or something. It was the worst. It took a while before my body felt like it belonged to me again." — Jenny

Shoe Shock

Jumping into puberty feet-first can be trying. Keep your sense of humor intact and know that your feet will be in proportion to your body in very little time.

"By the time I was in seventh grade, I noticed that my feet were growing. I found myself tripping all the time. I kept outgrowing my shoes. It seems like it ought to be fun going shoe shopping, but I got to where I didn't like it. I just wanted my feet to stop growing." —Rachel

The Great Race

Growing up isn't a race. Whether you experience changes sooner or later, you'll find that it all evens out in the end.

"I first noticed that the girls around me were changing when I started seeing how tall they were getting. I was shorter, so I felt left behind. I really, really wanted to grow and get taller. It kinda bugged me that my growth spurt came later than some of my friends' growth." —Becca

Changes, Changes

No doubt about it, it's not easy when growing up means dealing with several changes at once.

I was always taller than my classmates, but then I began seeing that my friends were gaining on me. That was kind of nice. I didn't feel so incredibly tall anymore. But then I started becoming more clumsy than I used to be. That was kind of hard to deal with. —Ali

Slow and Steady

There's nothing fun about nailing yourself every time you're on the move. Take it easy and move a little more thoughtfully till you're feeling a little steadier.

"I was always catching myself on the corner of the coffee table at home. I'd bang my hips going through the doorway or crack my hand on the kitchen chair as I was getting up. I was covered with bruises all the time, and I felt like the biggest doof." —Alex

Breakout Bummer

No one likes to see her complexion take a trip to zitville. Keep your chin up if this happens to you and check out skin savers that can help you save face!

"I've always been tall, and so I didn't really notice that I was getting taller. But what I started seeing was that my skin was breaking out. It started getting slowly but surely worse." — Emma

Perspiration Situation

No need to sweat this situation. There are all kinds of reek remedies at hand that can help you put a stop to smell.

"During sixth grade, I began noticing that it reeked in the gym locker room after gym class. Some of the p.e. teachers started complaining, and they told all of us to start putting on deodorant. Good thing. Everyone was perspiring way more than before, and so was I." — Jenny

Early Bird

It might feel like it's hard to fade into the background when breasts start taking center stage. Keep in mind that others who notice may secretly envy you and hope they start developing soon, too.

"I'd always had a flat chest, but one day I noticed I was developing. I was pretty early compared to my friends. It was weird to start getting breasts before anyone else did. People seemed to really notice, but I didn't care that much." — Jemma

Cool Curves

Look on the bright side when your body starts taking shape!

"One day, I started noticing that my hips were becoming curvier. At first I wasn't sure if I liked it. But right away, I realized that my pants fit better. I began seeing that getting curves was a good thing. I could wear some of the latest styles now that I couldn't wear before when I was stick-straight. I secretly liked the fact that my body was now getting some shape." — Trish

Puberty

These girls are describing the first visible changes that they noticed. Changes like these are often the first signs of puberty, the period of your life when your body starts to shift and transform. Chemicals in your body called hormones stimulate physical development. So there really is something you can blame for all the strange things going on with your body. Hormones are the culprit!

The whole reason hormones kick into gear is that they're helping prepare the body for reproduction. What's reproduction? Simply speaking, it's the process of making babies. That's right. The whole purpose of puberty is to prepare a woman's body for growing a baby.

Whoa, you may be thinking. *I have to go through a whole bunch of changes when I'm still in elementary or middle school? I have to go into an emotional Tilt-A-Whirl now just to be ready for the someday possibility that I might have a baby?*

True—you're a long way away from the time you'll be having a baby. Still, your body is going ahead and getting itself prepared well ahead of schedule. That's just how Mother Nature works. And like it or not, you're going to have to go along with it.

Puberty Alert!

Signs that you're entering puberty might include (in no particular order):

- ✿ Rapid growth spurt
- ✿ Arms and legs growing visibly longer
- ✿ Feet increasing in size
- ✿ Becoming somewhat uncoordinated
- ✿ Appearance of breast buds (see Chapter 5)
- ✿ Hair growing in places it hasn't before
- ✿ Appearance of skin blemishes
- ✿ Increase in perspiration
- ✿ Vaginal discharge

Many girls discover that changes like these occur during their fifth- or sixth-grade year. This is generally when you are ten or eleven years old. Of course, your changes can start to occur before this or even a little later. Some of them could start even before you're aware of them. And you might find that you experience some of these on a slightly different timetable than your mom or your older sister did.

No doubt about it, having your body transform almost before your eyes can be a little weird. It can seem like a huge deal and can take up a lot of your attention. You might find that you need some extra time to get used to all the new things happening to you. You might find that one day you're fine with it, and then the next, you're freaked.

As you're going through this transition, be extra kind to yourself. It's easy to get down on yourself when so much is going on at once. Don't get caught in the trap of putting yourself down because you're not sure of what you're feeling. It doesn't do any good, and in fact will make you feel worse as you go along.

The Putdown Trap

How do you know if you've fallen into the putdown trap? Do you ever hear yourself saying things like:

- ❀ "I'm such a dork."
- ❀ "I wish I looked like her."
- ❀ "No one will ever like me because I look the way I do."
- ❀ "I'm a total geek."
- ❀ "What an unbelievable loser I am."
- ❀ "I'm waaay behind everyone else."
- ❀ "Something must be insanely wrong with me."
- ❀ "Why would anyone want to be friends with me? I'm so out of it."

If any of these thoughts are running around in your head, you're being way too hard on yourself—and totally unfair!

It's easy to start sending yourself messages of self-doubt when you're feeling mixed up about what you're going through. But it's just as easy to be aware of what you're doing and avoid it altogether. So, the next time you're about to give yourself a mental slap, make a point to send an upbeat message instead.

Avoid the Putdown Trap

Here are some quick fixes for getting over putdowns.

Coolness Quotient

Remind yourself that you really are cool, and that your coolness quotient isn't going to change even though your body does. Being cool has nothing to do with starting your period or your breast size or how many curves you have. Cool is being first and foremost your own fabulous self. End of story!

Un-funnies

Avoid constantly making jokes at your own expense in front of your friends. It's not funny to treat yourself as a less-than person, and it can make your friends feel uncomfortable as well. It won't make you feel better if you constantly try to seek reassurance from your friends by joking in this manner.

Fab Femme

Look at yourself in the mirror and tell that great girl just what it is that makes her such a fabulous femme! Mention your plus points, such as "You,_____(say your name), are caring, compassionate, and you go the extra mile to try to be helpful; you work hard at school and at being a good family member; you strive to stay healthy; and you have big dreams." There. That ought to get you started. Now keep going. Don't hold back on the self-praise when you're in the privacy of your own room. It's not bragging; it's self-affirming.

Someone Else's Shoes

If you still have trouble with this idea, treat yourself as though you are your own best friend. You probably can't imagine telling your best friend that she's a horrible person. Use your imagination and say things to yourself that you'd say to a superclose buddy.

See the Good

Make it a point to view every new thing that happens to you as something positive. At the very least, look at each bodily change as proof positive that you're healthy and on your way to becoming a woman. Even if you don't totally understand it all right now, try to see each thing that happens as a new challenge and a new discovery.

Keep in mind that while you're going through changes, so are your friends. Be easy on your buds. After all, they're just as weirded out as you are about the things that are happening with their bodies— even if they try to cover up their insecure feelings. Just because you can't read their minds doesn't mean they aren't feeling a bit out of whack themselves.

So, even if you want to crack jokes at your friends' expense to make yourself feel better, DON'T. You'll only make them feel terrible, and you'll feel even worse about yourself knowing that you're not being a very caring friend.

Instead, when they worry (and maybe even whine) about something bodywise, step in there and be SuperFriend. All girls are in this puberty thing together—the best thing you can do is love and support one another. So if your best buds seem a little grumpy or down, don't panic. Do your best to help them use the rules in the Avoid the Putdown Trap section to get back on track.

By now, you've probably figured out that your mind and body are closely connected. That is, if you take care of your emotions, you'll be on your way to dealing with body changes in a more positive way.

Now it's time to read on and learn more about your body beautiful. In the next chapter you'll look a little more closely at what you can expect when puberty steps onto your personal stage. Just keep in mind that changes don't occur in any particular order. Some girls experience several of them at once, while others find they happen one after another. It all depends upon your unique makeup.

4

The Scoop on Puberty: What's Going on Here?

As you've already learned, puberty is the name given to the process of changing that you will be going through. It's the combination of things that happen to your body when you hit your teen years.

Zoom in for a closer look, and check out the head-to-toe changes you'll be facing in your future.

Growing, Growing, and GROWING!

The growth spurt usually is one of the first tipoffs that your body's heading into the puberty zone. Sure, until now, you probably cruised along growing at the rate of about two inches a year. Suddenly, you find that you're growing at about twice that rate! While maybe you're thrilled that you're adding some inches, you might be a little uncomfortable with these new changes. Like, maybe you're not so psyched about the fact that your favorite formerly long T-shirt is now exposing your bellybutton. And it's not because it shrank in the wash! Not to worry. Your growth will slow down in less than a year. You'll probably stop growing within three years of starting your period.

Or maybe you're on the opposite side of things—all of your friends are shooting up and you're not growing at all! Don't panic. Your day will come. And remember—no matter what heights you reach bodywise, there is no limit to what you can do! You already have tons of special talents, and as you grow up you'll discover even more. And that's more important than height any day!

Girl's-Eye View: Growth and Girls

Find that you're outgrowing the boys? That's because boys tend to mature later than girls do and often start their growth spurt a few years later (girls start puberty on an average of two years earlier). So if the guys tease you about your height, bear in mind that most of 'em probably wish they could lose their short-stuff status, too!

Arms and Legs—The Long and Short of 'Em

Starting to notice that your sweatshirt sleeves are going from full-length to three-quarter length? Feeling like your pants are hiking up past your ankles at an alarming rate? You think, "Maybe the washer and drying are shrinking my clothes."

Nope. Your clothes aren't getting smaller—your arms and legs are actually growing extra long during this time. You might look at yourself in the mirror and see a girl who looks so much like a colt with long, gangly legs that you feel like whinnying! If so, don't start picturing yourself getting ready to run in the Kentucky Derby just yet. Your arms and legs will end up being in proportion to the rest of your body, and you'll lose that coltish look.

In the meantime, however, you might notice some aches here and there, especially in the leg region. Some people refer to these as "growing pains." These aches are caused by rapid growth of bones and muscles. They generally aren't really painful, just annoying. But if they do ache to the point where you're beginning to think growing is a real pain, you might want to check in with your doctor for treatments that can help ease the ouch.

For example, fourteen-year-old Alex found that she was growing so fast, her knees began bothering her. Because she was a competitive cross-country runner, at first she thought her aches were due to her running activity. But when she went to her doctor, he evaluated her and determined that much of her discomfort was due to her rapid growth. Her doctor was able to give her appropriate treatments to help minimize her discomfort until her growth slowed down.

Feet Frenzy

As if your height, your arms, and your legs weren't enough . . . your feet are growing, too! You outgrow your fave shoes. So you buy a new pair, and before you know it, you've outgrown them, too! Will they ever stop, you wonder, or will you end up with basketball-player-sized feet? No need to fret. Your feet tend to grow for about a year. When they stop, they'll be in proportion to your height (whew!).

The Klutz Factor

With so many growth changes, you're bound to feel a little strange.

Like, maybe you used to be fairly coordinated, but now you're tripping over your feet at the worst times. Where before you could manage an obstacle course in gym class with ease, you suddenly find you can hardly even walk through the house without bumping into the furniture. You zing around a corner and bang your hip on the door frame. You constantly catch yourself on the corner of your desk at school, or hook your foot on a table leg.

With all of these mishaps, you're beginning to wonder if your friends are going to sell tickets to your next spectacular klutz performance. You'll probably notice a few more bumps and bruises, as well.

Puberty: When It's Not a Team Player

If you're an athletic girl, you might find that all of these growth changes are getting in the way of your competitive progress. For example, if you're a swimmer, you might find that your swim times slow for a while. Your strokes might be off. If you're a rhythmic gymnast, you might notice changes in your flexibility, or that your timing's just not right. These changes will work themselves out as you adapt and learn slightly different ways to go about your goals.

So are you doomed to be athletically impaired forever? No way! You'll soon be comfortable in your "new" body. Just be ready to adjust slightly (talk with your coach about your concerns), keep on playin', and wait patiently for the day when your body's back to its winning ways.

Body Shape Changes

You already know that girls' bodies come in all shapes and sizes. By looking around at school, for example, you can see that some girls are:

- Naturally long and lean, with long, slender necks and long, thin fingers
- Big and muscular, built along athletic lines
- Short and compact, built along stockier lines

When you glance at your body in the mirror, you can see that you too have a specific body type. But then—*wham!*—just when you thought you knew what your body looked like, suddenly you see it's begun to change. Where once you had straight lines, you now see the beginning of curves. Along with breast development, you might also see that your hips are starting to get some pads of fat and become curvier as well. As you start developing, your waist may seem a little thicker at first, but as you continue to develop, your waist may begin to curve in and become more defined.

Danger? Curves Ahead

When you see your body changing and filling out you may start worrying. Maybe you feel unsure as your slim, straight body begins to become curvier and more womanly. You might make the mistake of looking at superskinny models and movie stars and think that's how you're supposed to look. (Uh, NO!)

Sometimes you become concerned about your weight. It could be that in the process of growing, you actually put on a bit of weight. You might take the fact that your hips are widening and rounding and your waist is thickening as a sign that you're overweight.

It's easy to forget that not everyone is on the same timeline when it comes to taking shape! You might start worrying if you compare yourself to friends who haven't undergone changes yet, or buddies who happen to have different body types than you have. You might get it stuck in your mind that you can go back to being stick-straight if only you'd lose a few pounds. And you might get the bonehead idea that you'd better start dieting right away. Maybe you decide you'll stop eating breakfast or lunch or that you'll start picking at your dinner. Maybe you think that you should eat just one kind of food or another so that you'll look like your former self.

BAD BODY MOVE! Don't even think like that! Remember, part of the reason you want to learn about growing up and changing is so you'll make smart decisions about your body and be as healthy as you possibly can be. Messing around by eating only certain foods or cutting out meals is going in the opposite direction of health. It's downright unhealthy—especially for a growing girl like you.

If you have concerns about your weight, never decide to diet on your own. There's no need to risk harming your health. Talk with your doctor to find out what her opinion is. Because she knows that there's a wide range of what's considered normal when it comes to height and weight, she may determine that you don't need to be concerned at all. And if she does think you could

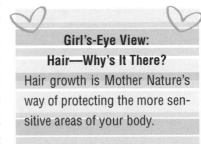

**Girl's-Eye View:
Hair—Why's It There?**
Hair growth is Mother Nature's way of protecting the more sensitive areas of your body.

benefit by adjusting your weight, she can suggest healthful ways to do so. What's more, she can monitor your progress so you'll be in the best shape ever.

When it comes to body shape, there's no right and wrong. The most important body type is the healthy type!

Hair—Why Is It There?

The hair on your head is sometimes called your "crowning glory," and for good reason: It's pretty and enhances your looks. But hair is also a protection provided by Mother Nature. The same holds true of the hair you grow in other places when puberty hits: It's nature's way of adding protection to delicate areas.

So when you start seeing hair growing under your arms and between your legs, though you might not like the way it looks, remember that biologically speaking, it serves a purpose.

When you are a little girl, your underarms are smooth. But soon, if not already, you may begin to notice hair starting to grow under your arms. This hair probably will be darker and thicker than the hair on your head. You can see it when you wear sleeveless tops or bathing suits. It's around this time that you might also begin to develop body odor. You'll read a little more about that soon.

Now, you may not pay any attention to this new hair growth, or you may not like the way it looks at all. Some girls decide quickly that they want to get rid of it, while others decide to wait a while. It's a pretty personal decision, and it's up to you and your mom.

You might find that removing this hair will help keep underarm odors under control. If you decide to want to give underarm hair a sendoff, check out the following section.

Fuzz-Free: The Facts

There are lots of ways to whisk away unwanted underarm hair: shaving, waxing, mittens, hair removal systems, and using depilatories. Whoa! Big word, right? Read on to find out more). After you read about these methods, you'll probably want to talk with your mom so the two of you can decide which one is best for you.

Shaving

Shaving removes hair at skin level using a razor. These days, you can find razors that are disposable and designed especially for women. These make it easier than ever to reach places with dips and curves, like the underarm area.

Some razors also have aloe vera strips on them to help keep skin from drying out and becoming irritated. This can really help when it comes to shaving the sensitive skin in the underarm area.

Razor Rave: The Lowdown on Shaving

Use these tips to help you decide whether shaving is right for you.

The Pluses: Shaving's cheap and painless.
The Minuses: Hair returns very quickly because it grows quickly, usually within a couple of days.

What to Do:

✿ To get a smooth shave, start by moistening your skin. It's probably best to begin shaving after you've been in the bath or shower for a while. Warm water will soften the hairs, making them easier to remove.

✿ Apply some shaving cream or gel to make the task easier. Give yourself ample time and use a soft touch. If you rush, you'll probably end up nicking yourself (uh, no thanks!).

✿ When you're finished, air-dry your razor. (Trying to rub it dry with a towel will only dull the blade and possibly ruin the towel.) Replace blades as soon as they become dull. Dull blades can nick and irritate the skin, making for a very uncomfortable shave!

Waxing

Waxing is a method that uses warm wax to remove hair. Because waxing has a major *ouch* factor—you've got to pull the wax from your skin quickly, much like you would a Band-Aid—it's not the most painless method, but you might like the supersmooth result.

Wax Watch: What Wax Can Do for You

Is waxing right for you? Read on to find out.

The Pluses: Waxing has the best long-term effects, because it pulls hair out from the root. After waxing, hair grows back in approximately three to four weeks

The Minuses: Um, can you say "ouch"? Also, you have to let the hair grow out quite a bit before you can wax again, which means having to put up with troublesome stubble.

What to Do:

✿ First, be sure to ask your mom if waxing's okay. You might want to ask for her help, too!

✿ Buy hair removal wax at your favorite drugstore. Follow the manufacturer's instructions. The instructions generally involve warming thin strips of wax, applying them, waiting till they cool, and then pulling.

Mittens

Hair removal mittens are another way to remove unwanted hair. They work kind of like sandpaper, and they're made of abrasive material. You wear the mitten on your hand and rub the skin surface to remove hair.

Mitten Must-Haves: How Mittens Work

Check out the info below to find out if mittens are right for you.

The Pluses: They're inexpensive and you don't have to worry about nicks and cuts.

The Minuses: They can leave skin abrasions.

What to Do:

❀ Read the manufacturer's instructions carefully.

❀ Basically, the instructions involve slipping a mitten on your hand and then very gently rubbing the mitten over the skin surface in a circular motion. Use it on a very small area at first, especially if you have sensitive skin, to make sure it won't cause irritation.

Depilatories

There's that big word again! Depilatories are foams, gels, creams, or lotions that dissolve hair chemically below the skin's surface. You apply the cream, wait for a while, and then wipe the cream and dissolved hair away.

Depilatory Decision: Are They Right for You?

This is a fast and painless way to remove body hair.

The Pluses: Depilatories are a painless way to remove hair (as long as you aren't allergic to them) and the hair stays away longer than it does if you shave.

The Minuses: Talk about stink! These chemicals can really put your nose in a tizz! Some manufacturers try to use fragrances to cover the smell, but you can still detect the strong odor. Worse, though, is that quite a few girls are allergic to the chemicals.

What to Do:

✿ Before trying depilatories, perform the manufacturer's suggest patch test.

✿ If you experience *any* symptoms of allergies (for example, redness, swelling or itching), DON'T use the product.

✿ If you show no signs of being allergic, smooth on the product after your shower. It helps to apply it after a shower because that's when the hair is standing up.

✿ Leave the product on for the manufacturer's suggested time (usually about four to twelve minutes). Remove it with a washcloth.

Hair Removal Systems

These hand-held systems look sort of like an electric shaver and are generally battery operated. They have rotating pads that remove hair.

Hair Removal Hints: What You Should Know

Is this system right for your bod? Read on to find out.

The Pluses: The manufacturers claim that these systems are painless and the hair stays away for up to two weeks.

The Minuses: They are more expensive than other methods (sometimes around $40). The appliance is bigger than razors or jars of depilatory or wax, so it's a little less portable if space is a consideration for you.

What to Do:

✿ Read the manufacturer's instructions carefully.

✿ Generally, you turn on the machine and run it over the skin surface.

✿ The rotating pads on the head of the device will remove the hair as you go along.

Pubic Hair: On the Grow

When you were a small child, the area on your lower torso, between your legs, was hairless or had very light hairs. But somewhere between the ages of eight and eleven, you'll start to notice a thicker type of hair growing there. Gradually, this hair becomes thicker yet until it covers this part of your body, which is known as the pubic area. The hair here generally grows in an upside-down triangle shape.

And, know what? Even if you're a blonde or redhead, you probably will still find that the hair growing in the pubic area is darker and curlier than the hair on your head.

More Hair—Where?

And don't be surprised if the hair on your legs and even your forearms becomes darker. Sometimes, girls may find that hair darkens on their upper lip as well. If you're bothered by this, you might want to talk with your mom about bleaching products made especially for this sensitive area. They can help lighten the hair and make it less noticeable.

Hair beware: Just don't try to bleach facial hair using hair coloring products that are made for the hair on your head. Your facial area is much too sensitive for regular hair color products. Be sure to only use the products made just for facial hair.

The Skin Story

Up to now, the changes you're going through haven't been that bad, right? Okay, so your body is growing and changing more than ever. Maybe you're even wearing a bra, but mostly things have been fine.

"Okay," you say to yourself. "If this is what growing up's all about, I can deal." But then, wham! Just when you start to care just a little bit more about how you look, your skin begins to change. And this isn't always in a way that you want it to be!! Enter the skin skirmishes.

When your skin meets up with puberty, things can definitely get a little crazy!

What's Happening to My Skin?

As you go through puberty, you'll find that your skin will change. For one thing, you might notice that your skin is starting to become oilier. For example, when you leave for school in the A.M., your freshly scrubbed face feels pretty good. But by midmorning, you begin to see more than a clean glow; you begin noticing an oily sheen! And this sheen only gets worse as you go about your day. By noon, your cheeks, your forehead, your nose, and your chin can really start shining. Talk about an oil slick! If you wipe your face with a facial tissue, you might even see a greasy stain.

Pimple Probs

As if the oil wasn't annoying enough, it can also be accompanied by unsightly red, raised bumps called pimples, or zits. Pimples can show up on your forehead, chin, cheeks, and nose. Talk about ruining your day!

The problem with pimples is that they can flare up at any time. And they always seem to pop up just when they're least welcome— before a big school event, a party, or a dance. Breakouts can be a scattering of smaller spots or a real rash of bigger, deeper bumps. Some of these bumps form whiteheads. Sometimes these bumps are accompanied by clogged pores, which appear as black dots. These are known as blackheads.

Sometimes pimples can leave scars. Sometimes they build up pressure under the surface and can actually kinda hurt. It's enough to make a girl cry!

You might be a lucky girl who doesn't have to deal with breakouts frequently. But you also may find that you have to pay serious attention to skin issues in order to keep breakouts under control.

Common Pimple Causes

What causes pimples and blackheads? Puberty is to blame. During this time, your body begins producing more oil, caused by increased output from the oil glands. This oil can cause clogged pores, which form blackheads. When this area becomes inflamed and reddened, this results in a pimple.

Pimples, unfortunately, can pop up on your face, back, or chest. So plan on devoting a little more time to skincare so you can keep a step ahead of pesky pimples. Make sure you use soap formulated for your face. Think twice before you grab the deodorant soap bar that happens to be in the shower to use on your face. And don't even think about using body bars with fragrances and dyes in them! These ingredients can aggravate skin problems and make them worse.

Instead, choose cleaning products that are labeled noncomedogenic. This means that they are formulated especially so they don't contribute to conditions that can cause pimples.

What Skin Are You In? Know Your Skin Type

Determining your skin type can help you determine what you need when you select skincare products. Here's how to figure out what skin you're in:

Oily skin has a noticeable sheen. If you blot tissue paper onto your face, you will probably see oil on the tissue.

Combination skin has oily areas—usually on the forehead and nose, called the T-zone. Cheeks and chin may appear to be dry.

Dry skin has dry, flaky areas, but can still be prone to breakouts.

Sensitive skin has a tendency to react to various chemicals, fragrances, and dyes in skin products. Might require hypoallergenic products (products formulated not to aggravate allergies) or even prescriptions from a dermatologist (skin doctor).

Acne Fighters

Okay, here's the good news: You don't have to live with a freaked-out face. It's often fairly easy to put a stop to pimples and other skin uglies. You can get yourself back in the clear zone simply by taking extra care to cleanse properly. Stepping up your skincare is a great way to start. Be aware of the products you're using to wash your face. Be sure to cleanse regularly with noncomedogenic products. If you still can't seem to put the brakes on breakouts, you might need to look for anti-acne ammo. Lots of stores carry products that can help you battle breakouts. They come in many different forms:

- ❀ Facial washes
- ❀ Soaps
- ❀ Facial wipes
- ❀ Lotions
- ❀ Cleansers

Look for products that are made especially for your skin type. As you scan the shelves, you'll probably see that there are all kinds of face-friendly acne fighters made just for teens. Take a moment to examine the different brands. Check out each product's specific purpose. Look carefully at the ingredients on each package. Choose products with zit-zapping ingredients such as salicylic acid or benzoyl peroxide.

If you use a skincare product with benzoyl peroxide, start with the lowest percentage (probably about 5 percent). You want to be sure your skin can deal with benzoyl peroxide first. If your skin doesn't react negatively to this ingredient, you can try it for a period of time to find out if this helps you battle breakouts effectively. If your breakouts continue, you might need to try something a little stronger. In this case, you might want to consider using 10 percent benzoyl peroxide. Make it a point to keep acne fighters away from your eyes. Some can really be irritating to eyes, causing burning and puffiness. Who needs that?

Skin TLC: Use Your Products Wisely

When you've found a product that works for you, use it as directed. Be gentle with your skin even if you don't feel like it's treating you fairly these days. Follow some of these easy hints:

- ❀ Apply zit-zappers with clean fingertips or clean cotton balls. You don't want to add to the bacteria count on your face, that's for sure.
- ❀ Be sure to cover the affected areas thoroughly and rinse completely, if directions call for rinsing.
- ❀ Keep in mind that overscrubbing your skin and applying harsh

products won't help your face clear up. In fact, all this harsh treatment can cause flare-ups to get bigger, to redden—and to look even worse than before!

Remember that you're always better off treating the skin you're in with some TLC, especially when it's giving you some trouble.

Makeup

If your mom agrees that you can wear light makeup or a little coverup to help you visually minimize skin hassles, look for products that work with your skin type and make sure they match your skin color. Choosing a color that's too dark for your skin could give it an orangey glow, so be sure to color-test makeup before buying it. To do this, use the color testers available in the store or ask a knowledgeable sales associate for help.

Don't think, however, that once you choose a color, you're set. Be ready to change skin-toned products seasonally because your skin color might change depending upon your sun exposure. The last thing you need is to look like you're wearing a mask because your coverup is a different shade than your skin. Most of all, check to make sure that the products are labeled nonallergenic or hypoallergenic and noncomedogenic.

Some Skincare Do's and Don'ts

Here are some do's and don'ts for maintaining healthy skin:

DO select noncomedogenic soaps made especially for your skin type.

DON'T scrub too vigorously thinking you'll put a stop to pimples. You'll only aggravate your skin further.

DO keep your fingers off your face. Scratching and picking can lead to infection and to scarring.

DON'T cradle the phone with your face, and avoid cupping your chin in your hand. Bacteria from your phone and hands can cause breakouts on your cheek and chin.

DO make it a point to use only fresh, clean makeup if you use makeup. Toss out old or dirty products. It's better to go without than to wear yucky makeup that might add to your skin woes.

DON'T borrow friends' makeup. It's a definite no-no. You might end up with skin or eye infections and aggravated skin problems.

DO take extra time to develop a skincare routine that works for you.

DON'T expect instant miracles.

DO follow directions and give the products a chance to work.

DON'T suffer in silence! If you find that your skin really gives you a tough time, even though you're trying everything to keep it clear, go see a dermatologist. These days, there are so many new techniques for tackling skin troubles—and your dermo can determine which kinds of treatments can make your skin look and feel better than ever!

Perspiration Problems

Find that these days you're sweatin' it more than ever? Puberty strikes again! Along with the increased production of oil comes the increased production of sweat from the sweat glands. This is more accurately called perspiration. Perspiration is often accompanied by an unpleasant odor.

"Oh, great," you might be muttering to yourself. "Growing up stinks—in more ways than one!"

Nope, not if you take steps to put perspiration in its place and stop the smell. To combat odor, take care to bathe properly each day and wear clean clothes. Try layering so you can remove a layer or two as the day heats up. Not only will you feel more comfortable, of course, but you'll also help keep perspiration to a minimum.

You might want to select clothing with looser, more breathable fabrics, such as cotton or cotton blends. Sometimes, synthetic fabrics such as polyester or polyester blends can trap perspiration, creating a moist environment. This is the perfect breeding ground for bacteria that can create odors.

If you find that you've entered the stink zone, avoid applying perfumes or colognes in an effort to cover the smell. This just won't do any good. And anyway, the combo of perspiration and perfume is, in a word, obnoxious! You might want to duck into a nearby restroom and apply a little soap and water to your underarms so you can feel fresher.

The Lowdown on Deodorants and Antiperspirants

Once you start noticing that your perspiration has an odor, you might want to start wearing deodorant or antiperspirant. What's the difference between the two?

Deodorants contain chemicals that mask odors.

Antiperspirant contains chemicals that slow or blocks the production of perspiration.

Check with your mom to see if she thinks it might be time for you to try one of these products.

Deodorants and antiperspirants are applied under the arms, usually in the morning after you bathe. Some girls reapply after sports or simply when they want to feel fresh again. Just be sure to apply them to clean skin. It doesn't do any good to apply deodorants after you're perspiring.

There are many types of deodorant and antiperspirant products available in drugstores, grocery stores, and convenience stores. They come in lots of different forms:

- ✿ Roll-on
- ✿ Solid stick
- ✿ Gel
- ✿ Spray

They also come in varying strengths, from mild to super strength. Some are scented with light fragrances. Some are unscented. If you have sensitive skin, you may be allergic to the chemicals contained in deodorants and antiperspirants. Before using new products, be sure to read the instructions provided by the manufacturers and test the products in the manner they suggest. Try a few different forms and brands so you can decide which one works best for you.

How can you tell if you're allergic to a product? Check out these signs:

- Redness
- Bumps
- Itching
- Burning
- Swelling

If you see any signs that you might be allergic, discontinue use of the product immediately. In cases like these, you might need to see a doctor for a recommendation.

In addition, if you try different store brands and find that none of them stops your wetness problem, you should also talk to your health-care professional. Some people find that they need special prescription products to help them keep perspiration under control.

Hair Hassles

Just when you thought you'd had enough of dealing with the results of increased oil and perspiration production, you begin to see that there's more to the growing-up game. That's right—you'll see the unsightly signs of oil and perspiration in your hair, too!

Could be that you'll find that your hair gets oily more quickly than before. You might see that your locks get limp and seem to separate. Your bangs might start to clump together. Maybe your hair looks not exactly shiny, but more like *slick* after a few hours of activity.

If this is happening to your hair, don't despair! Just know that there's no getting around it: You simply have to take care of your crowning glory a little differently than before.

Wash-a-Rama

If your hair is too greasy or oily, try washing your hair more often. If you used to wash every other day, try washing every day instead. You also can try keeping shampoo and conditioner handy in your gym locker if you shower after gym class.

Brushing

If you only shampoo once a day, keep brushing to a minimum. Extra brushing stimulates oil production. So slow it down by avoiding brushing your hair unnecessarily.

If you cut down on unnecessary brushing, don't make the mistake of running your fingers through your hair to "comb" it instead. Dirt and oil from your fingertips can accumulate on the strands, causing just the oil slick you were trying to avoid!

Shampoo-U

A word here on shampoo: Use enough to clean your hair thoroughly, but don't think you need to go for a double dose. Read the instructions on the bottle to make sure you're using the right amount. There's no need to waste shampoo by using twice the amount you need. If you do the job right the first time, you don't need to repeat.

In addition, while you're lathering up, don't think you need to scrub extra-hard to remove oil. Be gentle to your hair and use your fingertips, not your nails. Your poor scalp doesn't need any scratch marks. And what's more, too much vigorous rubbing will bring on extra oil, which is the last thing you want. So take it easy!

Girl's-Eye View: Hair Play

Are you a hair twirler? In class, do you sit at your desk twirling away while the teacher talks? If you constantly twiddle and twist your hair, the oil from your fingers contributes to your tress troubles!

Shampoo Switch-a-Roo

Sometimes a little change can do you good. While the same old shampoo might be fine for the other members of your family, that doesn't mean it's best for your changing hair. It might be time to do a little personal shampoo shopping. Select specific hair products that suit your particular hair needs. Look for shampoos and conditioners made especially for oily hair. Use them according to directions, and make sure to rinse thoroughly. Effective products aren't necessarily the most expensive products. They are simply the ones that are right for you.

Hair and Skin Harmony

Keep in mind that good haircare and good skincare go hand-in-hand. Lifting sweaty bangs off your forehead can help keep breakouts at bay. Consider hairstyles that keep your hair off your face to help keep your face in the clear.

When your body's changing, it makes sense to change the way you care for yourself from head to toe!

Quick Quiz

Time out! Before you read on, test your girl wisdom by taking this true-false quiz.

1. Yikes! Growing up can be confusing and sometimes more than a little scary. True False

2. Making fun of yourself can help you feel better about the changes you're going through. True False

3. It helps to arm yourself with information before your body starts undergoing changes. True False

4. If your friends start changing before you do, that means something's wrong with you. True False

5. It's not good to be different. It's better to try to look like other girls. True False

6. It's a good idea to talk with trusted adults about personal body stuff, even if it's soooo embarrassing. True False

7. It's easy for everyone to talk about body changes that concern you. True False

8. It's okay to tease your friends if they're worried about some of the crazy changes they're going through. After all, it makes you feel better about your own worries. True False

9. Your breast size isn't a measure of your self-worth. True False

10. The best way to handle puberty is to keep taking care of yourself the way you always have rather than making any changes. True False

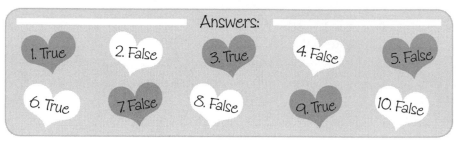

Answers:

1. True   2. False   3. True   4. False   5. False

6. True   7. False   8. False   9. True   10. False

Scoring

If you scored between:

8–10 right   YOU GO, GIRL

You've got a great attitude about the growing-up thing. You realize that it's not a race to get there first and that everyone changes in her own way. You know you might not be totally sure of yourself at all times, but you're willing to learn so you can make the most of your experience.

6–7 right   YOU'RE GETTING THERE

Though you make it a point to gather information, there might be some moments when your uncertain feelings get in the way. You might feel so overwhelmed at times, you kinda shut down, hoping that becoming a woman is something you can ignore and it'll go away. Keep in mind that there are lots of reasons to feel fabulous about growing up. It's all in your attitude.

0–5 right   KEEP TRYING!

Body confidence? What body confidence? Okay, so you're just not comfortable with the idea yet that growing and changing is all part of being a girl. You might be quick to kick into close-your-ears-and-close-your-mouth mode, which makes it hard to ask questions and listen to the answers! So consider keeping an open mind and paying attention to the positive signs that your body's changing for the better!

5

Breast Development: When, Where, Why, and How

So when do your breasts begin to make their grand entrance? For some girls, breast development starts at age eight or nine. For others, it may not happen until twelve or thirteen, or even as late as sixteen If your breasts do not develop at all, you should talk to your mom. No matter when your breasts begin to grow, just remember—there's no right or wrong time for breasts to begin to grow. You might feel a little weird when it finally begins to happen. After all, it feels strange to start experiencing something that seems so. . . well, grown-up.

How Does It Work?

Breast development begins when estrogen (a female hormone) is produced at increased levels. Chances are, when you're about eight or slightly older, you'll notice that your chest is no longer flat. The dark, round area called the areola that's around your nipple starts to push out, forming a sort of bump. This is called a breast bud. It'll be more sensitive than usual for a while. The areola will also grow larger. You might even notice some darker hairs growing around it.

Don't be surprised if you have only one breast bud for a while—this is totally normal. Though it's noticeable to you, chances are that no one else notices that you're uneven at all. The other breast bud will appear before long.

As you continue the process of breast development, you'll notice that your nipple and areola stick out from the breast area. But gradually, the area will merge back into the breast.

Once your breasts begin to develop, you'll see that the area around the nipple begins to swell. This tells you that milk glands are developing. These milk glands enable a woman to nurse her baby. The emergence of these milk glands signals that your body is readying itself for this future task.

Your breasts may develop quickly, or they may develop slowly. All girls are individual and special, so it all depends on your body.

So What's the Right Size Anyway?

About the time your breasts put in an appearance, you might start noticing just how, um, *interested* people are when it comes to the subject of breasts. Maybe they're more than simply interested. They're like . . . *obsessed*. Your friends might compare their growth with other girls. They might talk about it a lot more than makes you comfortable. After all, up till now, chests haven't been a big deal. And they are a part of the body that people consider pretty personal and private. Yet suddenly you see it's become The Topic at your school. You've probably heard (or said) stuff like this in your classes or in the halls:

- ❀ "Can you believe so-and-so is wearing a bra?"
- ❀ "Amy's kind of flat, doncha think?"
- ❀ "Wow. Megan sure got big over the summer."

You've probably noticed that there are all kinds of words to describe breasts, words like "boobs" and "grapefruits" and others that aren't really cool to say at all. That's because our society seems to pay quite a lot of attention to breasts. No matter what size your breasts are, remember, they are part of what makes you uniquely, specially, eternally you. And part of being you is being proud of yourself—whether you're a D cup or an AA. But, as the old saying goes, the grass is always greener on the other side. If your breasts are large, you probably wish they weren't *sooo* noticeable. And if you have small breasts, you probably wish they were a little *more* noticeable.

So what's better? Large or small? Check out the following sections and see what you think!

Large . . .

All you have to do is turn on the tube or flip through a magazine to see that lots of women in the media have fairly large breasts. Newsflash: Some of these breasts are computer enhanced! That's right—they're made to look larger than they actually are. But this doesn't mean big breasts are the only kind there are. It also doesn't mean that having big breasts is best.

In fact, some girls with large breasts wish theirs weren't so large. They hunch their shoulders forward in an effort to downplay their breast size, or maybe they wear loose-fitting clothing to cover them up. Sometimes competitive swimmers wear two suits to try to downplay their sizable chests. Maybe you even do some of these things!

So, even though TV, movies, and magazines may make a big deal about it, you can see that some people don't think big breasts are the biggest deal ever.

. . . or Small?

While some girls are trying to cover up their breasts, other girls desperately wish they were bigger. They may wear padded bras, or stuff their bras with tissue paper. The opposite of girls who are trying to cover up, they might wear tighter clothing to play up their chests and give the illusion of increased size. They may even try to eat foods that they heard will make their breasts grow. (Myth alert: There are no foods that will make your breasts grow!) The point is, however, that you should never think it's, well, *less than* to have smaller breasts. As tempting as it may be, don't fall into this trap! Judging your worth by physical traits is never a smart game to play. Instead, be true to yourself, be true to your friends, and never forget that you have girl power—that's the coolest power in the world!

The "Right" Answer

Get it: There is no one type of breast that's "correct." Repeat it. All women are different and special, and so are you! So what's normal? You are! No matter what size your breasts are, you are beautiful. Don't worry about what's "normal." Normal doesn't always mean looking like everyone else!

Some women have big breasts, others have small breasts, and still others have medium-sized breasts. Some women have large nipples and areolas on their breasts; others have small ones. Some women have coffee-colored nipples, and others have rosy-colored ones. You might see that your nipples stick out a lot or that your nipples are inverted—that is, they are set back into the areola. It's not all about size—breasts come in different shapes as well.

The right kind of breast is the kind you have, plain and simple.

Girl Gab: Breast Talk

Don't worry! You're not the only one out there with breast concerns. Check out what these girls had to say about their own breast development.

Worry Wart

Crissy worried a bit at first when it seemed like other girls were "blossoming" before her:

> When I was in sixth grade, it seemed as if everyone was suddenly starting to wear bras. I was surprised. I had never thought about things like this before. Lots of the girls were wearing sports bras. Some of them were beginning to get bigger chests. I wasn't developing at all. I felt like the only girl who wasn't. At first, it made me feel like something was wrong with me.

Jemma had the opposite worry:

> "I was the first of the girls in my class to develop. It was weird to be going through something like this when no one else had much of a chest yet. Because I was the first, people kind of noticed. The boys sometimes

said stuff. It was annoying when they made wisecracks, but really, I didn't care that much."

Lopsided for Life?

If you have the same problem as Jenny—no need to worry! Here's what she had to say:

"When I first started developing, one of my breasts became bigger than the other. I knew no one else could really tell, but still was pretty upset about it. I'd heard in health class that this sometimes happened. Yet I kinda worried that maybe I'd be the exception and that I'd be lopsided for the rest of my life. Turns out, I didn't need to worry. The other breast caught up right away."

Bosom Buddy

Is your best bud as kind as Emma's? Make sure you're always being the best friend you can be!

"My breasts are kind of small. I was talking about it one day with a friend in sign-language class, and I told her that it bugged me to be smaller than most of the other girls. She reassured me though by saying that I shouldn't care about being small, that there are just so many more important things than the size of your breasts. I know she's right. I'm sure my time to blossom will come anyway. I just kind of hope it comes soon."

Slow Starter

Some girls are slower to start than others, but Katherine did her best not to let it bug her.

"I didn't develop until way after my friends did. At first, it didn't really bother me. I guess it was because I was busy with other things and didn't have time to really think about it. But then after a while, I wondered why I wasn't growing at all."

Cover Up

Alex tried to hide her breast growth—but that's not always the most comfortable option!

"I started developing and didn't really like the idea that this was happening to me. So I wore these loose shirts hoping no one would notice. But one day, my best friend looked at me when we were changing before a race and said, "You need a bra." I absolutely wanted to die. But I went out and bought one and felt better after that. I could ditch the loose shirts and just wear normal clothes again."

Your First Bra? No Big Deal

Like Becca, you might not be in a hurry to follow the herd. Remember, do what's comfortable for you.

I saw that the other girls were starting to wear bras, but it didn't make me want to rush out to get one. I just kind of didn't think about it. When my mom suggested we go shop, I shrugged and went along with it without giving it much thought.

Posture Perfect

Rachel learned that by hunching to hide her breasts, she was hiding something else—her confidence!

"I felt like I was kinda big-breasted, and when I first developed I became pretty self-conscious about it. I kinda slouched anyway because I felt so tall. Soon I started crossing my arms in front of my chest. I made a point to carry my books in front of me — anything to hide my chest.

"Then I started taking modeling classes, which made me become more aware of how I was carrying myself. I realized that I wasn't fooling anyone, and anyway, it looked worse to slouch than to just stand straight and be proud of who I was. One good thing that came out of my experience is that I became more comfortable with my looks. In the end, I think that's all that matters!"

Starring . . . You and Your Breasts (Oh, no!)

If you're starting to develop in a big way, you might start noticing that, um, well, all eyes are on you. Or, more specifically, on your chest. Your friends might start saying things. And maybe it's not just your friends. Your mom—whom you've always trusted—starts making embarrassing announcements at the dinner table . . . and even in public! Even though you know that your friends and family don't mean to hurt your feelings, all this unwanted attention might make you feel kinda whacked out.

If the talk, talk, talk about your blossoming breasts is getting to you, calmly tell whoever's running off at the mouth to cool it with the comments. Ask these folks to respect your privacy. If you stay cool and calm, they'll think twice before bringing it up again. And before you know it, their attention will be on something else. (Whew!)

Buying Your First Bra

As you hit the late elementary-school or the middle-school years, you might begin to feel as if everyone at your school is rushing out to buy a bra. Maybe you totally need to get one, too. Or . . . maybe you aren't sure just what the fuss is about.

Even if nothing's happening to you yet in the breast department, you still might think you want to get a bra like the other girls. If so, go ahead—talk with your mom and see whether she agrees that you're ready for a bra of your own.

Bra Battles

If you're bothered by your bra-less state, but Mom says it's not time yet, check out some of these options:

- ❀ Try wearing a tank top under your shirt
- ❀ Try wearing a camisole under your shirt
- ❀ Start with a sports bra that doesn't have distinctive cups

If you're feeling like you need something, but you're not quite ready for a bra, any of these options might help you feel more comfortable.

However it happens, soon the day will come when it's time to get your first bra. The next section will help you select the bra that's right for you.

Choosing the Right Bra

Wow, what an exciting moment! It's finally time to choose your first bra. When you're choosing a bra, you should have two main goals:

1. Pick a bra that fits you well.   2. Pick a bra that's comfy to wear.

Before you head out to the store, you might want to flip through some teen magazines and look at bra advertisements. This can be a fun way to "window shop" in private and get to know what kinds of styles are out there. Still, when it comes to actually buying one, there's no substitute for trying one. After all, the most important part of buying a bra is making sure it fits. Plan on taking time to make a careful selection. Buying your first bra shouldn't be a grab-n-go experience. Finding a bra that works for you takes a little more than closing your eyes and snagging the first one off the rack. You should plan to take enough shopping time to get the right fit and the right bra for you. You definitely need to try before you buy!

Your bra should feel good and look smooth under your clothing. But how can you be sure to find the right one? That's easy!

A bra that fits . . .

- ❀ Doesn't bind or pinch or cut into your skin.
- ❀ Doesn't leave red marks when you take it off.
- ❀ Doesn't sag or bag.
- ❀ Feels comfortable whether you're standing or sitting (or even running!).
- ❀ Moves with you.
- ❀ Doesn't hike up when you lift up your arms.
- ❀ Doesn't bunch up noticeably anywhere under your clothes.
- ❀ Doesn't come unfastened as you go about your day.
- ❀ Has straps that stay on your shoulders without falling down!

When and Where?

But when and where should you buy your bra? It's really up to you and your mom—and what you're comfortable with.

When you arrive at the lingerie section of your favorite store, look for the teen or juniors section. If your store doesn't have a special section set aside, don't worry. Most lingerie sections still have sizes to fit teens, even if they're mixed in with bras for older women.

If you want to ask a saleswoman for help, go for it. If you're squeamish about a stranger helping you with this task, remind yourself that this is the salesperson's job. She's not there to judge you—she's simply looking at you as a customer. And when it comes to bras, she knows her stuff. She can help you make selections more quickly and can share her knowledge with you as well. Still, it's your decision. If you are too shy to seek help, chances are you and your mom can search for bras on your own just fine.

Take a moment to flip through the racks of bras and familiarize yourself with different kinds of bras. Pull them out and examine them. These days, there are so many to choose from in all kinds of fabrics and styles. While you're at it, you might as well know what's available, even if not all of the bras you see suit your style. If you're shy about the whole bra thing, choose a time when you know the store will be less crowded. This way, you can browse without feeling like the eyes of other shoppers are on you.

Even if you're not shy, you might want to shop when the store isn't populated with people. You'll probably feel more relaxed and more inclined to take the time needed to make good decisions.

Okay, so all of this instruction might make you think that finding a bra that fits is hard to do. Really, though, it's easier than it sounds. Read on!

Types of Bras

Bras come in all kinds of price ranges. They also come in a zillion shapes, styles, and fabrics, so you're sure to find one that meets your special needs. The following list explains the types of bras you'll most likely see when you're out browsing.

Back-Hook Bras

True to their name, these bras hook in the back with one or more little metal or plastic hooks. They take a little reaching around to clasp, but most girls manage this just fine.

Front-Hook Bras

As you would expect, these bras hook in the front. Some girls prefer this for ease of fastening.

Strapless Bras

These bras stay on without the use of straps. They're good to have if you like to wear tube tops. They also come in handy for the day when you wear a strapless dress. However, strapless bras sometimes aren't as comfortable as regular bras, so you might want to save them for when you have special clothing needs.

Underwire Bras

These bras have special wires that curve under your breasts for added support. The wires are sewn into the fabric. If you have larger breasts, you might find that underwire bras are a comfortable choice.

Silicone "Bras"

Not exactly bras, these plastic cuplike objects are probably best for older bra wearers. They add shape and form to your breasts and stick right to your skin.

Padded Bras

These bras have thicker fabric cups to give the appearance of having bigger breasts. Padded bras can be very tempting to girls whose breasts are slow to develop, but you'll probably find that you're more comfortable without the extra padding. Remember—you're beautiful just the way you are! So, if you're tempted, try sticking with what Mother Nature gave you first—be strong, be confident, and be yourself. No padding required!

Bras are generally made of cotton, but they come in other fabrics as well. Some of the more colorful bras are made of colorfast polyester. Some bras are made of stretchy materials, like Lycra or spandex so they can move and bend with you.

Bras have all kinds of interesting names. They might promise to make you look fuller and rounder and all-out fabulous. Of course no bra is going to change your life and make you somebody you're not. But that doesn't mean you can't look at them all!

Your Type of Bra

So which is right for you? Try them all on and determine which works best for you.

Beginning Buds

If you're just starting to develop breasts, look for a training bra. No—a training bra doesn't train your breasts to grow or anything. It's just a term that describes a starter bra—one that comes in small sizes with cup sizes ranging from A to AAA. Okay, so this bra might not look like some of the more grown-up bras, but it's a great way to get used to the whole wearing-a-bra deal.

Curvy Chick

If you're a well-endowed girl with bigger breasts, look for a bra that gives extra support. This will help keep your breasts from bouncing when you run—and all of that bouncing can make you feel pretty uncomfortable. You might want to check out the many bras that are made with thicker straps and strong, supportive fabrics. Underwire bras might be a good bet also, since underwires are specially placed to give your breasts a little extra support.

Girly Girl

Try looking for a lacy bra that brings out your feminine side. You might even pick out one that has a little embroidered detail like a rose or a pearl front-and-center for an added girly touch! Keep in mind, however, that these lacy bras are pretty delicate and need extra care. Even if you like the look of lace, you should also select a few practical bras that can stand lots of washing for everyday wear. Keep in mind that lace bras might not look good under shirts made of thinner fabric, such as T-shirts.

Sporty Girl

Look for a sporty model that lets you play without interfering with your movement. Sports bras come with extra support so your breasts won't flop and jiggle while you're, say, trying to kick a soccer ball toward the goal or guiding your favorite horse along a mountain trail. Sports bras are also cut in different ways to allow for greater range of movement, so you can raise your arms to set a volleyball or swing a tennis racket. These bras are usually made of strong, breathable fabric that keeps you cool and dry and stands up to lots of washing.

Trendy Girl

Look for bras that come in some of your favorite colors or styles. These bras might have funky designs on them, or extra trim. Okay, so trendy bras won't make a bit of difference to your breasts. But they can make growing up a little more fun and make you feel just a little more playful as well.

No-Nonsense Girl

If you're a no-frills girl, check out simple, white or neutral-colored cotton bras. These bras come without doodads like little bows and flowers, so they look smooth under your clothes.

No matter what type of bra suits your personality, have a good time with your new purchase. Who says you always have to be totally serious about your body changes? Sure, you can be practical, but why not also add a little fun to growing up?

There's More!

Okay, now you know what type of bra suits your personality, but that doesn't mean you're ready to buy! First, you have to get a sense of your size. There are two parts to your bra size:

1. Rib size   2. Cup size

So how do you get your rib size? It's sooo easy! You need to get a measuring tape and measure your chest width. You can get your mom to help you with this.

To get your rib size:

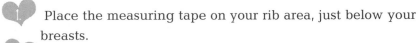

1. Place the measuring tape on your rib area, just below your breasts.

2. Wrap the tape around you so that it's comfortably loose but doesn't sag.

3. Read the number on the tape. That's your rib size.

Rib Size and Bra Size

Now let's translate rib size to bra size. This chart will show you how. If you're . . .

22–23 inches, the size to look for in a bra is 28

24–25 inches, the size to look for in a bra is 30

26–27 inches, the size to look for in a bra is 32

28–29 inches, the size to look for in a bra is 34

30–31 inches, the size to look for in a bra is 36

To get your cup size (which will be a letter):

1. Measure around the fullest part of your chest with your tape measure.

2. Subtract your bra size from this chest-center size. That will give you your cup size.

Now it's time to translate the measurement to a cup size. Find your cup size in the following chart.

If the difference is −1, your cup letter is AAA

If the difference is 0, your cup letter is AA

If the difference is 1, your cup letter is A

If the difference is 2, your cup letter is B

If the difference is 3, your cup letter is C

If the difference is 4, your cup letter is D

Once you know your bra size and your cup size, select a few bras that you like so that you have some choices.

"Fitness" Tips

When it comes to bras, keep one word in mind: comfort! That's what it's all about. The cutest little number is a no-go if it doesn't feel right.

Want a great fit for sure? Check out these comfy cues:

1. Fasten the bra and check to make sure it doesn't constrict your breathing. If the bra is too tight, unhook the clasp and rehook it at the next notch. If the bra is too loose, unhook the clasp and rehook it so that it's tighter.

2. Lean forward to let your breasts fill the cups.

3. Take a good look in the mirror. Turn and look at your profile. Move your arms a bit so you can see the sides of your bra. The sides should be level, not riding up.

4. Run your hand under the straps over your shoulders to make sure the straps aren't cutting into you. If they are, try adjusting the straps. Look for the plastic or metal piece on the straps that allows you to make them tighter or looser. If you adjust the straps and you find that they're still annoying, try another type of bra.

5. If the bra sags, put it back. Choose another bra that's either a smaller chest size or a smaller cup size.

6. When you feel like you've got a comfy fit, slip your shirt over your bra and see if you like how the bra looks under your clothes. If the bra seems smooth, you've probably got a winner.

7. Consider buying several bras that fit so that you can change them to suit the clothes you're wearing and so you have one to wear if another is in the wash.

First Bra Flashbacks

When it comes to first bras, nearly every girl has a story to tell. Check out some of these cool stories.

Dressing Room Drama

Crissy let her fears get the best of her . . . but you don't have to let them get the best of you!

"When I went to buy my first bra, I wanted to make sure no one I knew saw me shopping. I just wanted to get it over with. I wanted to just sink into the floor when my mom loudly asked this saleslady to help us. 'My daughter needs a bra,' she practically shouted. I couldn't even look at the saleslady. She kept handing me these bras to try on through the dressing room door. I was so afraid someone I knew would come into the store. I grabbed a few in several different sizes, pretended to try them on, and bought them without caring if they fit or not."

Freaked Firsts

If you're really worried about being seen bra shopping, perhaps you can follow in Jenny's footsteps and go to a different store or town.

My mom took me to buy my first bra. I insisted that she go to a store in the next town because I was so worried that someone would see me shopping for something so personal. I hoped I wouldn't bump into anyone I knew, but once inside the dressing room, I didn't worry so much anymore. I stayed in there forever trying to figure out which bra fit right. They felt weird enough as it was. I didn't want to have straps cutting into me. I also didn't want to buy something that looked too big and ridiculous on me.

Early Start

Jemma started early so she could get comfortable with wearing a bra.

"I started wearing bras in fourth grade, even before it was necessary. I think it helped me to do this because even though I developed earlier than my friends, I already felt comfortable wearing a bra."

Bra Pressure

Remember—do what's best for you. Just because everyone else is wearing one doesn't mean you have to. And yet, if it makes you feel better—like it did for Rachel—go ahead and get a head start on wearing a bra.

"I bought a sports bra even though I really didn't need one. I was tired of being practically the only girl at my school who wasn't wearing one. Anyway, I figured I might as well get used to wearing a bra. I was going to start developing some day, no matter what."

Hurry Horror

Because she was in a rush, Alex chose a bra that was sooo not right for her. Take your time when choosing your bra to be sure you get the most comfortable fit.

"I was in a big hurry to just leave the store, so I grabbed the first bra I could find. I tried it on, but didn't pay attention to the way it fit. Boy, was I sorry. Over the next few days, I was so uncomfortable because my bra was just too big. The cups kind of scrunched and looked kinda lumpy under my clothes. The straps kept slipping, and they drove me crazy. It was two weeks before I could go back to the store and get a bra that fit."

Surprise, Surprise

Katherine's mom planned ahead.

"I didn't even think about stuff like bras. But one day, I was getting ready for a costume party, and I was wearing white tennis clothing. My mom handed me this bra that she shopped for without telling me. I didn't really feel like I needed it, but there it was, so I just put it on. I just kind of fell into wearing bras after that. I guess my mom went out and bought the bra to make it easy on me."

Ali's Angst

Ali was a little concerned about wearing a real bra, but she adjusted quickly.

> I started wearing sports bras just, well, because. Then one day, my mom bought me a bra and gave it to me, telling me to wear it. I thought it was weird and didn't want to at first. But then I decided not to think about why. I just started wearing it. Pretty soon, I stopped being concerned about it at all.

Final Bra Thoughts

Okay, so now you've learned your bra size, cup size, and what type of bra best suits your body and your lifestyle. But there are a few more things you might want to know before moving on.

The First Few Days

The first days of wearing your bra, you might feel very aware that you have one on. You might even be worried that your friends and classmates know you're wearing one. But guess what? No one has x-ray vision, so you're totally safe. And even though it seems strange at first, it won't be long before you don't even notice it at all. Putting on your bra will be just a part of getting dressed—just something you do without giving it a second thought.

Girl's-Eye View: Bra Bit

Remember that you're constantly growing and changing. So be sure to check frequently to make sure your bra still fits. It's easy to get so used to a bra that you forget to notice that it stopped fitting months ago!

Caring for Your Bras

As you go about your busy day, you'll probably find that you can work up a real sweat. Don't forget that your bra is absorbing perspiration just like any other article of clothing, so be sure to wash your bras often. Since bras tend to be delicate garments, you might want to handwash them with gentle soap so that they'll last longer. Let bras air-dry, if possible, or set your dryer on low heat. If you choose to machine-wash bras, place them in net bags made for lingerie to keep them from snagging. Set the machine on the gentle cycle.

Bras at Night?

No way! There's no reason to wear a bra at night—no matter what silly myths you might have heard. Your breasts won't start sagging if a bra does not support them for a few hours. It won't affect your breast development in any way. In fact, wearing a bra at night might even feel confining and uncomfortable while you're sleeping. And because you should always be at your most comfy when you're getting your zzzs, plan on putting on your bra each morning and removing it each evening.

"Growing and changing doesn't have to be a scary thing that just happens to you and takes you by surprise. It can be much more manageable if you talk to others around you who have been through it before. Talk with your mom, your older sister, your aunt, or your caretaker. These women can calm your fears and give great advice. They can tell you stories that make you see that your concerns aren't way out there and weird."
— Jenny

6

Getting Personal: What's Going on Down There?

Okay, we've spent a lot of time talking about the changes you're going through on the outside of your body—growth spurts, body hair, and breast development. But there's more! There are tons of things happening right now "down there" —with your genitals and your reproductive organs—that you probably don't even realize are happening. This chapter will teach you about some of the things going on in your body that you can't see.

Your Genitals

The genitals are your sex organs, which enable you to have babies. Boys have genitals, and so do girls. You're probably already aware that boys' genitals are more visible than girls' (See Boys and Puberty later in this chapter), but just because your genitals are a bit more difficult to see doesn't mean you can't check them out.

Okay, so it may seem like a weird idea, but in order to find out more about yourself, it helps to take a closer look at your genitals. Can you imagine going through your whole life without knowing what you look like down there? Your genitals are uniquely, especially your own, and you were probably taught from the time you were little to keep this area private. That's a smart idea, but it's also a smart idea to take a moment to find out more about your body.

You can learn more about how your body looks by:

❁ Looking at diagrams. By studying the drawings, you'll get an idea of what your body looks like on the outside.

❁ Looking at your body using a mirror. When you look at your body using a mirror, be sure to pick a time when you're feeling comfortable and you have privacy. Flop back on your bed so you're comfy, and spread your legs. Use the mirror to take a look at yourself.

What Are You Seeing?

Congratulations! You're on your way to discovering new parts of yourself! This is the "you" that you've never seen before. In this section, you'll learn more about the different parts of your genitals.

The Vulva

As you look at yourself, you'll see the area where pubic hair may be starting to grow. On little girls, this area is flat. When you enter puberty, a small layer of fat builds up over this area, causing it to rise a bit. This entire area is called the *vulva*.

Labia Majora

Bring the mirror down, where you'll find two skin flaps, called the *labia majora*, which means "big lips." They are positioned on either side of a long, narrow line. They are smooth and farther apart on little girls, but as the girls enter puberty, they become fuller and seem to move closer together. This helps protect the highly sensitive area within.

If you're fairly far along in your development, you'll see that this outer set of lips is covered with pubic hair.

Labia Minora

If you hold back the labia majora, you'll see some inner labia. These are referred to as *labia minora*, or small lips. As with outer labia, these inner labia are smaller in little girls. But they too grow as girls enter puberty. Keep in mind that although these lips are called "small," they might actually be bigger than the outer lips. They can be pinkish or brownish in tone, depending on your body. They might have some ridges or be wrinkly in texture.

Vagina, Urethra, and Clitoris

By separating the inner lips, you can see three organs. The first one is your *vagina*, which is an opening into your body. The next one is located above your vagina. It's another opening, called the *urethra*. This is the one through which urine comes out.

At the top, you'll see the *clitoris*. This sensitive area looks like a little bulge. It's very small, but supersensitive, because it's jam-packed with nerve endings. If you touch it, you might notice that it's pretty tender.

The vagina is the opening that leads to the internal sex organs. It's got moist, elastic walls and resembles a small tube. It might be partially covered by a ring called the *hymen*. The hymen might be prominent, or it might be something you can hardly see at all. Perhaps it has some openings in it. A girl's hymen might have openings or tears caused by sexual activity or vigorous sports activity.

You can't see all the way up the vagina. But if you were to be able to, you'd see the *cervix*. At the center lies the *os*, or opening of the cervix. This is the opening through which menstrual fluid flows from the uterus and down through the vagina. This is also the opening through which a baby passes (again from the uterus and down through the vagina).

During puberty, your external organs will shift and change. Sometimes they will enlarge. Because these changes are visible, they tend to be the ones you're aware of. But these external organs and their changes are only part of the story. The real deal is that a whole lot is happening inside your body as well.

Inner Workings: The Female Reproductive Organs

Ready to take a closer look at what's going on internally?

It's quite possible that thinking about inner bodyworks isn't at the top of your Things To Do list. In fact, the whole subject might cause your eyes to glaze over. After all, the idea of learning about something you can't see might not seem that important. "Out of sight, out of mind," you might be thinking. Yet getting the inside story is a primo way to really understand why your body's behaving and looking the way it does. So gear up to get a closer look at the "inner" you!

Check out the list of organs below. Can you guess what they all have in common, other than that they're all inside of you?

- ❀ Ovary
- ❀ Ova
- ❀ Fallopian Tube
- ❀ Uterus

- ❀ Os
- ❀ Cervix
- ❀ Vagina
- ❀ Hymen

These organs all are connected with the process of having a baby. They're generally called the female reproductive organs.

The Ovaries and Ova

Let's start by searching for the *ovaries*. These are located on either side of the pelvis, and they are a sort of "storage center" for ova, which are female reproductive cells. Ova are also called eggs. Actually, the ovaries contain hundreds of thousands of eggs. As you enter puberty, your pituitary gland, which controls various bodily functions and is located in a depression of the skull's base, starts to produce chemicals that cause the eggs to begin to mature. The pituitary gland also produces hormones called estrogen and progesterone, which affect the reproductive organs. The matured egg leaves the ovary. This process is called "ovulation." During ovulation, some women feel a

Girl's-Eye View:
Inside Your Ovaries
Your ovaries contain hundreds of thousands of reproductive eggs, called ova. That's a lot of eggs!

twinge in the abdomen area. If an egg—an ova—were to meet up with a sperm, which comes from a male, the egg would be fertilized. This would result in a zygote, which would eventually develop into a baby.

Fallopian Tubes

Fallopian tubes are the channels through which the ova travel from the ovaries to the uterus. If you were to look closely at the fallopian tubes, you'd see tiny projections that look like fringe. These fringelike projections fan back and forth, gently guiding the egg from the ovary to the fallopian tube.

The Uterus

The *uterus* is a small, upside-down-pear–shaped organ located in the center of the pelvis. This is the place where a zygote, or fertilized egg, grows into a baby. The uterus is amazingly elastic. It can expand a great deal to make room for a baby. And once the baby's born, it will become small again.

How Big?

So, you're probably wondering how all of this important stuff can fit inside your body, right? Let's break it down to size. In a grown woman:

- ❀ One egg = approximately the size of a pinpoint
- ❀ Fallopian tube = approximately four inches long
- ❀ Uterus = the size of a plum (when a baby is growing inside a woman's body, the uterus stretches like a balloon so that it can fit the baby)
- ❀ Vagina = three to five inches long (it also stretches when a man's penis enters it, and stretches more to allow a baby to travel through when it's being born.)

The Cervix

At the other end of the uterus, you will find a small mass of flesh with a hole in the center called the *cervix*. The opening goes between the uterus and the vagina.

The vagina is a corridor that links the reproductive organs to the outside of the body. It is here that a male deposits his sperm inside a woman's body. Sperm cells have a role in fertilizing an egg. Semen is the fluid that contains the sperm. The sperm travel up the vagina, through the cervix, to the uterus, and into a fallopian tube. If an egg were present in the fallopian tube at this time, the sperm could fertilize the egg.

Menstrual blood comes out of the vaginal opening and exits the body. A baby also makes its way from the uterus out of a woman's body through the vaginal opening.

Baby girls have all of these reproductive organs when they are born. As girls grow, the organs grow larger.

Changing as You Grow

The reproductive organs grow as a girl approaches womanhood, and in addition, the organs may also shift their position.

The uterus, which in a little girl was positioned up and down, now usually leans forward or backward slightly. The bottoms of the fallopian tubes also shift forward. In some women, the uterus may stay straight up and down, while in others, the uterus actually tilts backward.

After the reproductive organs grow and shift position, they are ready for the development of a baby. When these organs are ready, they'll let you know. This is when you'll have your first period, which is called *menarche* (this is pronounced me-NAR-kee).

Boys and Puberty

Think girls get all the grief when it comes to growing and changing? Think again! Boys are going through puberty, too. Like girls, guys' bodies change and their emotions go on roller-coaster rides. And believe it—guys find that all this is weird and overwhelming at times, just like you do!

What Are Boys Going Through?

For starters, guys generally start going through puberty later than girls do. Whereas girls might start around age eight or nine, guys tend to start around age eleven. So right off the bat, they feel sort of left behind, even if they're not sure why. They look around and see girls starting to take off on their growth spurt. A guy can feel kinda goofy slow-dancing with a girl who's tall enough to pick him up and put him in her pocket. Okay, so that's an exaggeration. But you get the point.

With guys, as with girls, hormones get the puberty thing rolling. A hormone called *testosterone* causes the external genitals (including the penis and testicles) to grow. It's also responsible for hair growth, including facial hair. You'll probably notice this on guys' upper lips. When the growth comes in thick and dark, many boys start shaving.

Boys also experience hair growth in the underarm area. What's more, they grow hair on their chests and legs. They, too, grow pubic hair, which grows around the penis and scrotum (the pouch containing the testes).

When guys hit puberty, perspiration and oil production increase. They, like girls, can find that the locker room starts to feel like a real toxic cloud filled with the scent of sweat. Guys will find they need to bathe after rigorous activity. They'll also want to consider clothing made of fabrics that "breathe," and will want to use deodorant and antiperspirant to help keep odors and perspiration in check.

Another outward sign that a boy has entered puberty? His larynx starts to grow. This is the bump on the front of the neck that is commonly referred to as an Adam's apple. With this growth often comes something that can really freak out a guy. His voice can start to crack with no warning. If you wonder why some guys hesitate to speak up, it's not that they're trying to act cool. It could be that they're worried that their voices will crack or that they'll squeak midsentence. So rather than risk this embarrassing moment, some will choose to avoid talking whenever possible.

After the voice-cracking phase, boys' voices deepen. While the guys prefer this to the unpredictable nature of squeaking and cracking, they still might need a little time to get used to this change.

Boys Worry about Changes, Too!

Keep in mind that just as you began noticing that girls around you were changing, guys start noticing that their buddies are going through transitions. They, too, can fall into the trap of playing the comparison game. They look around and feel like something's wrong with them if they start going through puberty before their friends have. Like girls, they also can feel anxious if they haven't started changing and everyone else has. What's more, guys can be kinda tough on each other, ribbing each other in pretty harsh ways.

And that's too bad, because like girls, guys have different body types too.

Girl's-Eye View: All Shapes and Sizes

Maybe you've happened to notice that the guys in your class come in all shapes and sizes. Some are tall and some are short. Some are built solidly and appear to be gaining muscles. Others are slender and reedlike. It's easy to see that there's no right or wrong way for a guy to look, just as there's no right or wrong way for a girl to look.

Guys too are bombarded with messages from the media on how a guy ought to look. They see movies where the heroes are big, brawny guys with well-sculpted bodies. Ever looked at pictures of guys in movie magazines or sports magazines? Check out the guys' chiseled faces, square jaws, huge biceps, and massive chests. These guys hardly look real. (And for good reason: Some of them are computer enhanced just the way girls are computer enhanced in the magazines you've seen.) Yet the boys at your school can get the mistaken message that they too are somehow lacking if they don't resemble these perfect pictures.

Skin Skirmishes

Guys also deal with skin skirmishes, and for the same reason that girls do: hormones cause their faces to break out. Boys too can experience oily skin, pimples, enlarged pores, and blackheads. And they don't like facing this nasty fact any better than girls do. They want to have clear skin just like anyone else.

The good news is that they, too, can learn to care for their skin and buy skincare products that help them zap zits. Unlike girls, however, most guys don't feel comfortable using coverups. While guys act like they don't care if they have to deal with breakouts, be assured that many of them do care—in a big way. They just don't tend to talk about things like this with their buddies.

Growth Spurts, Mood Swings, and Other Stuff

When guys finally do start growing, their growth spurt can be pretty dramatic. They grow taller, but they also develop muscles.

Their shoulders and chests start to broaden. Their genitals grow as well. Their testicles enlarge and the penis grows longer and might darken in color as well.

While this is going on, some guys find that their genitals can ache. It's generally a minor discomfort, but it can be noticeable nonetheless. For the most part, guys don't feel comfortable discussing this with anyone, but it can affect their moods.

Guys get moody but tend to act on their feelings in slightly different ways than girls. Whereas girls might feel free to be teary and giggly at times, guys are more apt to clam up when they're feeling sensitive. Sometimes guys have trouble articulating their feelings and might even determine that they're angry when they're actually worried or feeling down. This can confuse the guys just as much as it confuses those around them.

And when it comes to becoming interested in the opposite sex, lots of guys are waaay later than girls. For this reason, you might notice that at first, the girls seem more interested in guys than the guys are in them. This is where it comes in handy to wait. The guys will catch up and start noticing girls. You just have to be patient!

Body Stuff Guys Worry About

Boys have plenty of worries—just like you. Check out this list to find out what boys are worrying about.

- ✿ Sprouting too much hair—or not enough hair.
- ✿ Having their voice crack and squeak just when they wish it wouldn't.
- ✿ Having one testicle bigger than the other.
- ✿ Having "wet dreams" (this when the penis releases semen during the night).
- ✿ Not being tall enough, or being overly tall.
- ✿ Not having well-defined muscles.
- ✿ Feeling their emotions swing up and down and not knowing what to do about it.

In the end, remember, guys are people too. While they grow and change in their own ways and on their own timetables, they are going through some confusing times as well. So when you're baffled by boy behavior, don't respond with a thoughtless wisecrack. Treat a guy just as you would treat a gal pal—with understanding and compassion.

Boys on the Brain

Puberty and boys. By now, you've probably clued in to the fact that the two are somehow related. You can't help but notice that at the same time that girls begin changing, boys start entering the picture.

You might prefer that this weren't the case. You probably wish this could stay a "girls only" subject. It's enough that you probably think you have to share this confusing condition with girls like you. But add boys to the mix? "Uh, no thanks," you might be saying. But there's no getting around it: growing and changing requires that you learn a little about how guys come into the puberty picture.

Boy Crazy?

Anyway, while all this body business is going on, you can't help but wonder what's up with your gal pals and how they are viewing the guy population these days. Where before they seemed to get grossed out by guys, now it seems that they can't get enough of them! You might be of a "Chicks rule!" mentality—and it might seem to you that all of a sudden your boy-crazed buddies care way too much about guys' opinions.

"When I started fifth grade, I started to see my friends start to drool over guys," Ali, now fifteen, recalls. "It was so weird to watch girls change like that."

Like Ali, you might see your friends paste up pictures of hunky guys in their lockers. Or maybe you catch your best bud doodling a guy's name in her notebooks.

Are your buds getting boy-blitzed before you are? You may be embarrassed that when some of your buddies pass a guy in the hall they don't just oh, so casually walk by the way they used to. Noooooo. Nu-uh! They stop. They screech. They stare. They giggle. They talk louder about absolutely nothing. Worse, they get all giddy! You, in the meantime, are more than mortified. You slink back against your locker and wish the ground would open up and swallow you. You want to signal the guy frantically and apologize on behalf of all girlhood for your friends' frenetic—and freaky—behavior!

Or—c'mon, it's confession time!—maybe it's *you* who's finding yourself thinking about guys more than ever. Before, yeah, you used to look at them merely as soccer teammates at times or maybe brotherly types, if you considered them at all. Yet lately, you can't help crushing on a boy who was formerly on your Most Obnoxious list. You catch yourself drooling when you happen to catch sight of him from across the schoolyard. You find yourself wishing that he'd notice you—as, well, *a girl!*—not simply as a mass of protoplasm that sits behind him in science class.

Hormone Happenings

Hello—what's up with all this guy goofiness?

You guessed it: Hormones once again wreaking havoc in your world. They're responsible for your changing feelings about guys and some of the confusion.

Add to this the fact that guys themselves are going through their own version of puberty, and it's not on the same timetable as girls. So they're just as confused as you are, which sometimes makes them act in ways that are less than predictable. In a word: baffling!

So if you're finding yourself in the crush zone, take things slowly while you venture into this new guy *terror*-tory. While feelings aren't right or wrong, you don't have to act on every feeling that you have just because it's there. Just because you think about Mr. Wonderful day and night doesn't mean you have to start stalking him around school. You don't have to become a pest. Just because you might even dream of kissing him doesn't mean you have to go for the Big Smooch at the first guy-girl party you might be invited to.

Plenty of people make the mistake of acting on impulse without engaging their perfectly good brains first. You might know someone who pestered a guy so much he blocked her IM. You might have a pal whom everyone's talking about because she goes on flirt overload every time she encounters a male of the species.

The thing is, many girls have crushes on guys they don't even know. They might have fuzzy feelings for a guy they've never even seen in real life, like a hottie on a TV show or in the movies. They might have the warmies for someone they've seen in real life but have never spoken to. This might be a guy they sit near in math class but have never even said hello to. If they've never exchanged two words, how can they be so sure that these feelings are so real?

Crush Clues

Here's the deal: Crushes, though they involve strong feelings, are not the same thing as love. They're a simply a sneaky substitute. Crushes crop up in part because your hormones are messing around with your mind. Love is something more serious that two people experience when they know each other well. So you can see that girls who take action based on crushes are bound to make some silly and potentially embarrassing moves.

That's not to say that crushes can't be fun if they're taken in the proper spirit. Have a good time with these new feelings, but be wise. Learn from your friends' mistakes so you don't have to feel foolish yourself. Consider your actions carefully so you can make good decisions. It's okay to wait for the day when you're more mature to act upon your feelings. In the meantime, behave appropriately around guys so you don't have to feel weird later. Chat with your mom about guys and about when she thinks the right time for you to date will be.

No matter what your mom's answer is, be sure to stay focused on your own healthy growth during your young teen years. There's so much great girl stuff up ahead, and you don't want to miss out on getting to know yourself better than ever!

"I believe that sometimes it's okay to do things ahead of time. I wore a bra way before I really needed to so that it wouldn't be a big deal by the time I actually did need one."
— Jemma

7

Your Period: The Next Big Step

he word "period" is short for menstrual period. This is the period of time each month when a fluid that includes blood flows from the uterus through the vagina, and then out of your body. This may seem strange and a little scary at first, but it's all perfectly normal and nothing to worry about!

When my period started, I was like, what is this? I had heard about periods before, but I was still confused. It's one thing to learn about it in class. But when it happens to you, it's suddenly all different.
— Jemma

Period FAQs

Check out these period FAQs to learn the whys and hows of this important event.

How Much?

Just how much fluid is involved?

Although you bleed for a few days and can soak a few pads during this time, don't worry. You don't actually lose that much blood. When all's said and done, it's generally about 4 to 6 tablespoons.

How Long?

How long does this fluid flow?

It varies from girl to girl. Some girls find that the flow lasts for three days. Others find it lasts as long as seven days (the average is two to seven days). Generally the first few days are the heaviest. Gradually, the flow diminishes down to a trickle and spotting before it stops altogether.

> I worried that I'd start in the middle of class or something.
> —Jenny

How Often?

How far apart are these periods?

On average, periods tend to occur about every twenty-eight days. However, the time between them can range from seventeen to forty-five days. The whole span of time from the first day of one period until the next one is called the menstrual cycle.

Keep in mind that when a girl first starts menstruating, the number of days between periods can be irregular. But after a while (up to three years), the number of days between periods becomes more regular.

Your Period and Your Brain

Most of the growing and changing associated with your period takes place in your head—more specifically, your brain. Don't believe it? Well, it's true. It starts in your *hypothalamus*, actually. The hypothalamus is the part of your brain that controls your menstrual cycle.

Your body contains chemicals called *hormones* that are distributed through your body. The pituitary gland regulates the flow of hormones. Different hormones have different jobs to do. Some hormones regulate your growth, and other hormones regulate your reproductive organs.

Specific hormones called estrogen and progesterone give your body the signal that it's time for breast development and puberty to begin. Once puberty is underway, these hormones will send out signals that regulate the menstrual process.

Your Cycle

Your cycle starts on the first day of your menstrual period. As you track your periods, refer to this as day one.

Day One

On the first day of your period, you may feel like you're running on low energy. This is because the level of estrogen within your body is low.

Day Five or Six

When your period ends, you might find your energy level starting to rise. This is because your estrogen level is up. The lining in your uterus will be building up at this time in preparation for receiving a fertilized egg that would be nourished in the uterus.

Day Fourteen or Fifteen

An egg matures and is released from the ovaries. The egg travels along one of the fallopian tubes and makes its way to the uterus. This trip generally takes four to six days.

The uterus has been preparing for the possibility that the egg might be fertilized. The uterine lining has been thickening because of the activity of estrogen and progesterone. This thickening menstrual material is actually blood-filled tissue that contains estrogen. The thickened uterine lining could support the early growth of a baby. Fertilization generally takes place in the fallopian tube. If a fertilized egg were to enter the uterus, it would attach itself to the lining and start to develop into a baby.

Usually, however, the egg that enters the uterus is not fertilized. If an unfertilized egg enters the uterus, the egg does not become attached to the uterine lining. Instead, it breaks down slowly and is absorbed.

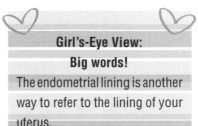

Girl's-Eye View:
Big words!
The endometrial lining is another way to refer to the lining of your uterus.

❝One of my friends started her period on the day our class went on this outdoor adventure program. It was a sort of wilderness survival experience. We were already a little jittery about having to spend a week being so rugged. My friend said it wasn't a big deal, but I was so worried that something like this would happen to me.❞ — Katherine

Day Twenty-Eight

When there is no baby to support, the thick buildup of menstrual material is not needed.

So, around day twenty-eight, if you do not become pregnant, the corpus luteum cyst will stop working, progesterone levels will fall, and the endometrial lining will become less stable. Since the thickened endometrial lining is not needed to help the embryo implant and growth, the material gradually sheds for a few days, flowing through the vagina and out of your body. The period of time when the blood-filled material is draining from your body is called the menstrual period. It can go on for anywhere from two to seven days

Generally, the first couple of days of the period is when the flow is the heaviest. It then begins to taper off.

"When Will I Get It?"

Okay, so it would be great to know *exactly* when you can expect your period.

It makes sense to want to know when to expect it. After all, who wants to be surprised when it comes to something like this? It would be nice to know in advance so you could prepare yourself. You could plan your outfits, plan your activities, and make sure you're somewhere where there's zero chance of getting into a sticky situation.

Unfortunately, it's not as simple as that. You really can't pinpoint the exact date you'll get your first period. Some girls experience their first period as early as age eight or nine, maybe even before the topic of menstruation has come up in health class. Some girls don't start till they're sixteen or even seventeen. The average age for a girl to get her period is twelve to thirteen. If you don't get your period by age sixteen, you should talk to your Mom and see a gynecologist.

Some girls start at about the age their mother did, or even at the age that their father's mother did. But don't get too excited—this isn't a foolproof method for determining when you'll start your period, either. Certain environmental factors come into play and you might find that their periods commence earlier or later than their mothers' did.

Elaine Pan, M.D., a gynecologist who often sees young patients, says that many girls today are experiencing menarche at younger ages than before. Why? Well, according to Dr. Pan, environmental factors, such as nutrition, might play a role in this. "Also," she adds, "girls who are obese might find their periods start earlier. This can be due to increased estrogen production in their bodies."

Will I Know When It's Coming?

You might not have any signs at all leading up to your first period. It might just happen—*wham!*—seemingly out of the blue.

On the other hand, you might experience an increased amount of vaginal discharge approximately six months before you get your first period. Some girls find that their periods begin approximately a year after their pubic hair starts to grow. A general rule of thumb? Your period might start approximately two years after your breasts begin developing.

When you first get your period, you may notice a feeling of dampness in the vaginal area. You'll then notice a stain on your underpants. This stain might be bright red, brownish red, or brown in color. There might be a lot of blood, or you might only notice a little bit of spotting.

I Got My Period! Now What?

Ooooookay, you say. It's official. The Big Moment has come. Now what?

It's totally natural to feel overwhelmed and a little freaked, even though you've learned what menstruation is all about. Remember, this is n-o-r-m-a-l.

At Home

If you're at home when you get your period, you can reach for the supplies you've stashed ahead of time in the bathroom cabinet. If you haven't stashed your own, tell your mom, and she'll be able to give you some pads. Check out Chapter 8 for all the information you need about the different kinds of supplies, how to use them, and which kind is best for you.

I used to worry that my period would start the night of the big school dance. In ninth grade, I looked forward to the big Winter Ball. It would be the first time I'd wear a formal gown, and I wanted everything to be perfect. But then, I'd start wondering if this could be The Night that I'd start my period. I'd picture myself ruining my dress and everyone staring. I kind of freaked myself out big-time over nothing. Of course, this didn't happen, and anyway, now I know it's easy to prepare yourself even if you started your period on a big night like this. — Rachel

At School

If you're in the school restroom when you discover you've started your period, here's what to do. Right away, you'll be able to see if you've bled a little or a lot. If you've bled a little, wipe away the blood, as much as you can, with toilet tissue. If you have a pad or tampon with you, you're ahead of the game.

If you don't have any supplies with you, fold some toilet tissue or facial tissue to place onto your underpants until you can get to the nurse for supplies.

If you've bled a lot, you might need to rinse out your underpants and spot-clean your outer clothing with cold water. Blot dry with paper towels.

Chances are, your menstrual flow will not have gotten onto your clothes. If it did, don't panic! Try tying a sweatshirt or sweater around your waist, if necessary. You might want to call your mom to see if she can possibly bring a change of clothes to school. If she's unable to do so, you might have to get a little more creative. Check out the ideas in the next section of this chapter—Girl Gab: In Case of Emergency.

In Class

If you're in class and have a suspicion that you've started your period, stay calm! Quietly ask the teacher for a pass to the restroom or the school nurse. Remember, the school nurse will understand your concern and be glad to give you a pad.

You don't need to feel weird about telling someone you trust that you need some supplies. Keep in mind that your mom, teachers, and school nurse have been through this. They know what it's like. Chances are, they will be happy to help.

Girl Gab: In Case of Emergency

Let's face it. Nearly every girl has an emergency at one time or another. No matter how hard you try to avoid it, your period might start unexpectedly and the flow might soak through your clothes. Fortunately, this doesn't happen often, and even if it does, it isn't as big a deal as you might think it would be. Read on for tons of tips from girls who've gone through this and know exactly how to handle it.

To Tuck or to Untuck?

Sometimes, you can use your own clothing for a quick coverup, as Trish suggests.

Untuck your shirt. If it's long, that can help disguise anything you're concerned about.

Use Your Ima-gym-ation

If you're lacking in spare duds to cover the evidence, Jenny has some really creative advice.

"Go to the gym locker room and see what clothing you can put together. You might be able to use your gym shorts under your clothing to replace wet underpants. If your school dress code allows it, you might be able to wear sweats to replace your outer pants. You might see if you can borrow something from a friend. Don't worry about looking like a fashion plate that day. All that matters is that you feel comfortable that nothing's showing."

Best Buds to the Rescue

What are best friends for? If all else fails, you can always snag a shirt from one of your buds, as Emma suggests.

"Borrow a sweatshirt and tie it around your waist."

Lost ... or Found?

Be an emergency fashion risk-taker. As Rachel suggests, the school lost-and-found could be your saving grace.

If you're really desperate and you don't have a sweatshirt or anything to cover up with, you might try going to the lost-and-found and borrowing something, anything. A jacket, a sweater, even a scarf that you can wrap around and pretend it's kind of a fashion thing. You can always wash what you borrowed at home and return it right away. But in the meantime, you will have saved yourself some embarrassment.

Skirt Switch-a-Roo

Alex has a quick fix for a period emergency that will get you out of trouble in a pinch.

"If you're wearing a skirt, turn it around if the back's stained. Then ditch your backpack in your locker and carry a stack of books low and in front of you until you can change clothes."

Girl's-Eye View: Just Ask!

If on the off-chance you find that the adult you turn to (like a teacher or school nurse) isn't helpful, look around for another appropriate adult. Remember that this isn't your fault and no one has the right to make you feel weird about a perfectly normal function.

It's Your Business!

When it happens to you, starting your period may seem like a big deal. And it is. But the real deal is that no one else besides your mom or whomever you choose to tell (like your best friend) has to know! There's no way anyone can tell by looking at you that you're wearing menstrual protection. If you use it properly no one has to know your private business except for you.

Period Firsts ... True Stories That Tell All

For some girls, starting their period is quite an occasion. Some cultures mark it as an important milestone in a girl's life. You might know someone whose family celebrates this event. You might not feel like celebrating such a personal event with anyone else. You might not want to celebrate at all. It's quite possible that you might even feel a little sad that you're not a little girl anymore. And you might feel a little grumpy about the prospect of having to be prepared from now on. Every girl reacts differently to her period. And that's okay.

In this section, our girl panel shares memories of their first periods. Every story is unique, just like every girl is unique.

Trish's Story

While at the time, Trish was a little bashful, she realized later what an important moment getting your period really is!

Trish happened to start her first period while on a minivacation in a desert resort town with her mom and a few of her friends and their moms. She wasn't exactly thrilled about getting her first period when she was away from home. But when her friends and their moms found out Trish had started, they immediately went into party mode.

"They made a big deal out of it. They were all happy, hugging me and dancing around all crazily. I was, like, 'What are you doing? Why are you so happy that I'm bleeding?' It was a long time later that I realized it was simply because they felt it was special that I had reached this milestone in my life. They thought it was cool that I was now a woman. Now I understand why they were celebrating."

Early Start

Kacie's period got off to a very early start, but she showed off her girl power and handled it like a pro.

"I got my first period when I was in third grade! I couldn't believe it. That was pretty early. Lots of my friends hadn't even started *thinking* about things like periods. Luckily for me, I didn't get too worried. That was because I'd heard about menstruating through a program at my place of worship. I also knew lots of older girls who had all been through it. Luckily, however, it happened at home. I just got some supplies from my mom's bathroom."

Late Arrival

Rachel proves that being last isn't so bad—and that talking about what's happening to your body can help you out later.

"I started way later than my friends. I was 15, a sophomore in high school. Up till then, I was a little worried that I hadn't started my period yet. After all, all of my friends had been having periods for years. But at the same time, I was secretly glad and felt a bit lucky that I didn't have to deal with it right away like most of my friends did. When it happened, I was kind of relieved. I felt that it meant nothing was wrong and that I was now menstruating like normal."

Worth a Thousand Words

Even though you may feel weird about talking to your mom about your period, Becca found out that her mom was able to lend a helping hand.

"A lot of my friends had started their periods in fifth and sixth grade. I heard them talking about it from time to time, so pretty soon I began feeling comfortable with the whole idea. I was kind of waiting for my own period to start. I mean, it just seemed like it was something that was going to happen to me pretty soon. I wasn't really worried. Mostly I just focused on what I would do when it happened.   (continued . . .)

I was twelve when I got my first period. I was at home, and I just got a pad and that was it. Later, I talked with my mom, and she gave me pointers on how to handle problems that might come up."

Period Panic

Ava shouldn't have had to go through her first period alone—if you're feeling concerned, *always* talk to your mom or another trusted adult.

"I grew up in a sort of rural area. People were pretty quiet about subjects like menstruating. When I got my first period, I was totally terrified. I had never talked to my mom about stuff like this, and I didn't know the first thing about it. I thought that I had cancer. I thought the reason for all the blood was that my insides had exploded. I was so afraid that I was going to die. I didn't talk to anyone about it for a long time. I just kept it all bottled up inside me."

Nothing to Fear

As Crissy learned, your first period is nothing to be afraid of!

"I was twelve when it happened. Before my first period, I worried a lot that it would happen at school and that I'd be embarrassed. I'd kind of imagine all kinds of embarrassing scenes for myself. It ended up that I started on a weekend when I was at home. The whole thing turned out to be no big deal."

False Alarms

You may have a few false starts before your period kicks into full gear. Check out what Emma has to say.

"I had a couple of false alarms, where I bled just a little bit. Like in sixth grade, I'd wake up and find a splotch of blood on my bed. I'd think, *This is it; I've started.* But then that would be it. This happened several times so that honestly I don't exactly remember when my first period really was."

First-Day Blues

Okay, so the timing of your period isn't always perfect. Jenny got off to a bumpy start, but it all worked out in the end.

"I was in ninth grade. It was my first day of high school. It was a brand-new place for me, and I was already nervous enough. I didn't know anyone at my new school. So when I ran to the restroom in the morning, imagine how bummed I was to see that I'd started my period. I'm glad I had stashed pads in my backpack. All day long, I felt like crying. It just seemed so unfair to have to deal with something like my first period on my very first day of high school. I mean, I already had so much on my mind!

It just so happened I had to have a medical evaluation to participate in school sports that very day. The [health-care professional] who evaluated me made me feel better by saying that it was a lot to deal with all at once and that I was being very cool about it."

First Period ... No Problem!

Alex greeted her first period by staying calm, cool, and collected.

"I started when I was fourteen. Sure, I knew what periods and all that stuff were because I'd talked with my mom about it. Besides, I'd taken health class at school, and we were given all kinds of info about it. But still, when it happened to me at school, I was surprised. Unfortunately, I wasn't prepared. But it honestly wasn't a problem. I just went to the office and told the nurse what had happened. She gave me some supplies right away, and I just went about my day as usual."

All Grown Up

Katherine was bummed when she got her period. While it's normal to feel a little sad, just think—you have so many exciting times ahead!

"I was pretty upset when I started my period. I was at home, and at first I thought, Whew. I'm lucky that it happened at home. But then I started to think about how this meant I was growing up. I got kind of upset because I didn't want anything to change. But there it was."

The Big Game

Ali's period came at a rather . . . *hectic* moment!

"A couple of my friends had started their periods by the time I was in sixth grade. Lots of my other friends hadn't started yet. My mom had talked with me about menstruating, and she'd given me this pamphlet to read. I knew the day would come when I started my period, but didn't really think about it.

The summer after sixth grade, I was playing on a community basketball team. During halftime, I ran to the restroom and saw that I had started my period. Even though I had expected it at any time, I really had never imagined it would happen right during the middle of a big game! I went, 'Oh, my gosh' and kind of freaked. But then I just bought a pad in the vending machine and took care of it. Then I went back out for the rest of the game. There wasn't anything else to do."

Your Monthly Planning

You've probably already figured out that it helps to be as prepared as possible when it comes to your period. By taking charge and being prepared, you'll know to have supplies on hand. You'll be able to be the boss of your emotions because you won't be taken by surprise. By feeling ready and in charge, you'll feel like a girl who's on top of the world.

When you first start getting your period, it probably won't come at the same time every month. At first, it might even skip several months at a time! What a pain, right? Well, your body is adjusting to all of the changes, and in about a year or so after your first period, you will probably be able to tell when it's about to start. Once this happens, you can start to understand your pattern by charting your menstrual cycles.

Dear Diary

If you like, you can keep a period diary that tracks the dates of your periods and how you're feeling. After a while, you'll probably notice that your body has a pattern.

As you keep your diary, write about the symptoms that occur before, during, and after your period, as well as how heavy or light your flow is each day. Jot down symptoms you might experience— moodiness, bloating, cramping, headaches, or clotting, for example— on the days they occur. (You'll learn more about these symptoms in Chapter 9.) By doing this, you'll begin to see that certain things tend to happen around certain times of the month. Knowing that they're related to your cycle might help ease any worry you might have about them. Writing about them also can help you vent a little along the way, so your feelings aren't all bottled up inside you. And that's a good thing about keeping a diary—it's a safe, nonjudgmental place to grumble and complain. Go ahead—let it all out! If you're having concerns, you can also use your diary to explain how you're feeling to your doctor.

Calendar Girl

If you're not into diaries but you love to be organized, try using a calendar to record the dates of your period.

To keep track of your cycle using the calendar, mark your calendar on the day you start your period. Write down each day whether your flow is light, medium, or heavy. And don't forget to mark the day your period ends. If you're not comfortable writing it out, you can even make up a kind of "period code" for your calendar.

Refer to your calendar so that you can plan the days that you'll need to bring supplies with you. You might choose to dress in a particular way that makes you feel

> **Girl's-Eye View:**
> **Stay in Charge!**
> Lots of girls find that keeping track helps them feel empowered and more in charge of their bodies.

more at ease. You might even want to plan activities according to the days you expect your period. For example, if your periods cause you to feel a little less energetic than usual, you might want to schedule something that requires top performance for a later date. Still, keep in mind that many girls are able to participate in high-level sports and perform at their best even though it's "that time of the month."

How Do I Know When It's Over?

Figuring out just when your period is over can be a little confusing at first. That's because you're not regular just yet. It might take a year or two before you are able to pinpoint just when your period will start each month and just when it will end.

In the meantime, you'll have to watch for signs that your period is about to end. Here are a few common signs:

* You won't have to change your pad or tampon as often. That's because the flow will begin to slow down.
* Your blood becomes more brownish in color, or the flow changes to a yellowish discharge.
* You might go for a day or so without any discharge at all.

Even if you don't see any discharge for a day or so, you should still wear a panty liner. You'll feel better knowing you have a little protection whether you need it or not. If you don't see any discharge for two days, you can generally be sure that your period is over.

But What Does It *Feel* Like?

Okay, so you're learning the when, where, and how of your period—but there's a little more to it than that. Want to know what having a period feels like? Check out what these girls have to say:

"It's kind of an achy feeling in your abdomen. It doesn't hurt; it's just a sensation of being heavier somehow." —Jenny

"It's a sort of damp feeling." —Rachel

"It doesn't hurt or anything. It just kind of feels like nothing." —Becca

"My stomach feels kind of tight on the first few days, but that's about all that I notice when I'm having my period." —Crissy

"I really don't feel anything at all." —Jemma

What's in a Name?

Menstruation—that's the official name of the monthly process of shedding blood and uterine material. But lots of girls use code words like these:

- ❀ The Blahs
- ❀ The Monthlies
- ❀ The Curse
- ❀ My Little Friend
- ❀ That Time of the Month

Whatever you call it, call it . . . just part of being a girl!

"Girls who worry about stuff should talk with their moms. You don't have to have one BIG talk all at once. Ask little questions all along the way, so that no one thing seems like such a huge, embarrassing thing. It's better to start asking your mom stuff rather than wait till she starts talking to you. That way, you can talk when you're in the mood. It's not as weird when it's stuff you want to know." —Ali

8

Swift Supplies: Stuff to Have on Hand

I f you've ever taken a stroll down the aisle that contains "feminine hygiene products," you probably couldn't believe just how many things there are to choose from!

Maxi pads!

Mini pads!

Panty liners!

Scented!

Unscented!

Tampons!

Slender!

Junior!

Absorbent!

Super absorbent!

Deodorant!

Ultra!

Extra long!

Gentle!

Looking at all of these items might make you feel a little, well, overwhelmed. It seems that there are so many decisions to make, and you don't know where to start.

Not to worry. All those colorfully packaged products can be broken down into two basic choices: pads or tampons.

Pads

Pads (also called sanitary napkins) are absorbent products that attach with an adhesive strip. They attach right onto your underpants. Pads are made of a soft, cushiony, cottony material. Some pads are lined on the bottom with plastic. This plastic helps to keep the menstrual flow from leaking through onto your underpants.

Maxi Pads and Mini Pads

You will find that pads come in different thicknesses. You can choose thick pads for days when your flow is heaviest. For the days when your flow is lighter, you can select thin pads. Some pads come with parts that stick out on the sides, called "wings." These wings can help to prevent leakage on the sides.

Some pads are labeled "overnight," which means that they are more absorbent and can be worn all night long. They generally are longer front to back to cut down on leaks when you're lying down.

Panty Liners

Panty liners are very thin pads with a tiny layer of absorbent cottony material and a plastic liner. Like sanitary napkins, these pads attach right to the crotch of your underpants. Some liners are labeled "thong," which means they are superskinny at one end for those who wear thong underwear.

Panty liners are useful products. They're perfect for the last few days of your period when your flow is next to nothing and you don't want to wear a pad. On light days, barely-there panty liners are much more comfortable than pads and keep your underwear from getting stained. They also can be used for extra insurance with tampons if it makes you feel more secure. (More about tampons later.)

Scented

These pads are treated with gentle fragrances to help fight odors. These odors can be caused by menstrual fluid and moisture trapped in the pads. Though these fragrances might help battle odors, they might also cause allergic reactions, such as bumps or skin irritation.

Unscented

These pads do not contain fragrances and might be better if you are allergy-prone or if you have sensitive skin.

Girl's-Eye View: Pads of the Past

Years ago, girls didn't have the convenience of pads with adhesive strips. When it came time for their period, they had to wear elastic belts with clips. These clips were placed at the front and back. Pads came with long ends that extended upward. The ends were bunched up and placed into the clips. While this got the job done, girls found that the pads shifted around somewhat, making leakage a lot more likely!

Pad Practice

Before your first period, practice putting on a pad. You can walk around for a while to get used to how it feels. You can make adjustments if you find you haven't pressed it on securely enough. You can check in the mirror and look at yourself from lots of angles, including the rear angle. This way, you can see for yourself that it doesn't show. So when the day comes that your first period starts, you'll be a lot calmer because you'll already be prepared. You'll already know that no one will be able to tell that anything is different about you that day if you don't want to tell anyone.

Tampons

Tampons are small, shaped kind of like crayons, and are worn internally. Like pads, they are made of cottony material. Some tampons also have rayon fibers in them. Rayon is made of cellulose, which is a natural fiber.

Unlike pads, tampons are specially designed to be placed into the vagina, and they have rounded tips for easy insertion. Like pads, tampons absorb the blood and tissue from your menstrual flow. Don't worry! Tampons don't plug up your menstrual flow. They absorb the blood and tissue. During heavy flow days, it's a good idea to change your tampon every few hours. As tampons become saturated, the menstrual fluid will leak out of the vagina. This is why you might want to wear a panty liner as a backup!

Tampons generally come with two types of applicators:

✿ Plastic applicators
✿ Cardboard applicators

Some tampons have no applicators at all (more on that soon!). All tampons have a string at one end that remains outside the vagina. You pull this string to remove the tampon after use. Some tampons are designed so that they can be inserted without applicators. These tampons are made for those who prefer not to deal with applicators.

Like pads, tampons come in different thicknesses. You can select a thicker, more absorbent tampon for the days when your flow is heavier. You can select smaller tampons for the days when your flow is lighter.

Girl's-Eye View: Pads of the Past

Girls who needed to purchase pads from vending machines found that they came with safety pins. These pins were used to hold the pads in place instead of a belt. Pins were certainly better than nothing, but once again were no substitute for the convenience of adhesive strips!

Pads or Tampons?

And now, for the big question: Which should you use—pads or tampons?

That's a decision each girl has to make for herself. Lots of girls prefer pads, while others think tampons are tops! Girls and women who prefer pads like the simplicity of using them. And some girls come from cultures that have beliefs about using pads over tampons.

Girls who use tampons tend to like the convenience they offer. They're small and easy to carry. Using tampons can cut down on odors. What's more, tampons allow girls to participate in sports such as swimming even while having their periods.

You'll probably want to talk about this decision with your mom so you can make the best choice for you.

Pad Pluses
❀ Comfortable
❀ Absorbent
❀ Get the job done

Pad Minuses
❀ Might feel a little bulky under light clothes on hot days
❀ Hard to tote around in a tiny purse

Tampon Pluses
❀ Convenient
❀ Small and easy to tote
❀ Worn internally, so you can swim while wearing it

Tampon Minuses
❀ Requires practice to learn to use
❀ Sometimes can be forgotten and left in your vagina, which can cause problems (see Tampon Trouble later in this chapter)

Like lots of girls, you might want to try tampons, but you feel kind of afraid. Maybe you've got some worries of your own. You might wonder:

. . . if tampons are hard to insert
. . . if it's, well, just plain, um, strange to have to touch yourself "down there" to insert them
. . . if it will hurt to put one in
. . . if it will feel freaky to have a tampon inside you

Girl Gab: Tampons

As usual, our girl group has lots of different feelings about using tampons. Check out what they had to say:

Tampon Trepidation

Jemma worried about the "yuck" factor. While tampons might not be the right choice for you, remember—there's nothing weird or gross about wearing them!

"Before I started using tampons, I didn't want to think about them at all. The whole idea was like, weird! Gross!"

For Older Girls Only?

Rachel was skeptical at first . . . she wasn't ready to wear tampons right away, so she waited until she felt comfortable.

I couldn't imagine using tampons at first. It just seemed much too complicated. Anyway, it was enough for me to get used to having to wear pads. Wearing tampons just seemed like it was something only older girls did. But soon I changed my mind.

To Learn or Not to Learn

Jenny was scared yet curious about the benefits of wearing a tampon.

"I was scared about using tampons. I worried that it would hurt putting them in or that maybe it would feel strange having something inside me. But I also half wanted to learn how to use them. I knew that they would make my life during menstrual times much easier. I could do things like swim without worrying what time of the month it was."

Take Your Time

Alex is taking her time to decide on the tampon issue—a very smart decision. Don't ever be afraid to take your time to decide whether tampons are right for you.

"I don't use tampons. No particular reason, I just haven't gotten around to learning how to use them. I guess I don't really need to know right now. It's just fine with me to use pads and I don't see the reason to have to change what I'm doing."

Menstrual Protection: Blast from the Past

Lucky you! These days, you can walk into a grocery store, drugstore, or convenience store and find all kinds of cool products designed to make your period easier to deal with. These products are brought to you courtesy of a combination of science and commercial enterprise.

But women who lived before the industrial age had to rely on Mother Nature. While Mother Nature did her best, you'll probably agree that today's products are more convenient. Check out these items that women used to use to absorb menstrual flow:

- ✿ Sea sponges
- ✿ Grasses
- ✿ Seaweed
- ✿ Animal pelts
- ✿ Dried moss
- ✿ Cotton rags

Most of these makeshift menstrual protection items definitely weren't convenient. They were bulky, uncomfortable, and fairly unsanitary. It wasn't easy to dispose of them, and it was even more difficult to scare them up during an emergency.

Whew! Aren't you glad you're a girl in today's day and age?

Pre-Period Practice

By practicing early, Emma felt more comfortable with tampons when the time was right.

> I tried to practice using tampons before my periods started. I'd swipe some of my sister's stash, but couldn't really figure out how to put them in so that they were comfortable. When I got my period for real, though, I got the hang of it, and I use them all the time.

Never Say Never

Some girls, like Katherine, aren't comfortable with using tampons at all at first.

"I used to think tampons were gross and that they'd probably be really uncomfortable. I used to say to myself, 'I'll never use those things.'"

As you can see, each girl has her own thoughts and opinions about tampons. No matter what you decide, make sure that you are comfortable with your decision and with yourself!

Tampon How-To

If you want to try tampons, follow these instructions for inserting a tampon with an applicator.

1. Buy tampons labeled "junior" or "slender." These are generally easier to start with than the thicker, super-sized tampons.

2. Choose the tampons that are the lowest absorbency for your level of flow. If you wear tampons that have a higher absorbency than you need, you might experience vaginal dryness.

3. Read the instructions provided by the manufacturer. Read them completely before you try to insert a tampon. Reread them if there are any parts you don't completely understand.

4. Wash your hands thoroughly so you don't introduce bacteria unnecessarily.

5. Sit on the toilet, or if you prefer, lie back on your bed, whichever makes you feel more comfortable.

6. Relax so that you'll find it easier to insert the tampon. If you're tense, you'll clamp your muscles, which will make it more difficult to insert the tampon properly.

7. Take off the paper or plastic wrapping and pull the string, so that it's hanging out of the applicator bottom.

8. Hold onto the bottom of the outside tube firmly with your thumb and middle finger and push the front end of the tube into your vagina.

9. When your fingers touch your body, gently push the smaller tube into the larger tube. Push up and slightly back. This smaller tube will guide the tampon into your vagina.

10. Pull both tubes out, and the tampon will remain in your vagina.

11. Wrap the applicator tubes in toilet tissue and deposit them in a lined wastebasket.

12. Wash your hands with warm water and antibacterial soap.

13. If you've inserted the tampon properly, you shouldn't feel any discomfort. In fact, you won't be able to tell at all that you're wearing one.

Certain tampons come without an applicator, and inserting them is a little bit different. The instructions below will tell you how to insert a tampon without an applicator.

1. Buy tampons labeled "junior" or "slender." These are easier to practice with than the thicker, super-sized tampons.

2. Carefully read the instructions provided by the manufacturer.

3. Wash your hands thoroughly with antibacterial soap.

4. Make yourself as comfortable as possible. Sit on the toilet, or if you prefer, lie back on your bed.

5. Relax so that you'll find it easier to insert the tampon.

6. Take off the paper, and hold the tampon firmly between your thumb and index finger.

7. Gently insert the tampon as far as it will go. Then nudge the tampon a little so that it will move up a little higher past the bony ridge. (Hint: If you can feel the tampon, chances are that you haven't gotten it in past the bony ridge.)

Girl's-Eye View: Keepin' It Clean

NEVER use tampons that have torn wrappers. If tampons are exposed to air, they can become dirty and covered in bacteria. It's never a good idea to introduce a contaminated product into your body. Check the wrapper carefully. When in doubt, throw it out!

Tampon Myths and Facts

With so much information floating around about tampons, you're bound to hear some crazy stuff. So what's real and what's not? Check out the myths and facts below to know for sure.

MYTH: Tampons are hard to learn how to use.

FACT: Not true! Lots of girls have learned for themselves just how easy they are. Sure, it might take a little practice, but it's just not that difficult.

MYTH: It's easy to hurt yourself using a tampon.

FACT: Not true! Tampons are soft and designed to fit easily into the vagina. Keep in mind that a tampon is smaller than the vagina. And don't forget that the vagina's walls are flexible. They can easily make room for a tampon.

MYTH: Young girls can't use tampons.

FACT: There's no reason why young girls can't use tampons. You don't have to be a certain age. If you're menstruating, you can use tampons.

MYTH: You can lose a tampon in your body.

FACT: Once again, not true! The strings are securely stitched. Anyway, tampons can only go up as far as your cervix. The cervical opening is much too small for a tampon to get through.

MYTH: If the string breaks, there's nothing you can do.

FACT: You can use your fingers to retrieve the tampon. In the rare case that you can't take it out using your fingers, you should see your doctor that day.

MYTH: Girls who haven't had sex (virgins) can't use tampons.

FACT: False again! Girls who haven't had sex usually have a hymen (or thin piece of skin) that goes across the vaginal opening. So it's easy to think that the hymen would have to be broken for a tampon to go in. However, menstrual fluid comes out through holes that are in the hymen—the same ones that a tampon could go through.

Tampon Trouble

Because tampons aren't visible like pads, you might forget that you've put in a tampon. Maybe you're just in a rush, or maybe the string gets pushed back so you don't have a quick reminder. If this happens, you might end up putting in another tampon, which pushes the first one further up the vagina.

If this happens to you, you might be able to reach up and extract the first tampon after taking out the second one. If you're not able to reach it, however, make sure you tell your mom and see your doctor right away so it can be removed. Sure, this might be a little embarrassing, but you risk serious health problems if you don't have it removed. Keep in mind that doctors have seen cases like this before, so you don't have to freak out if you find yourself in this situation.

Eco-Friendly Talk

What does wearing a tampon mean for Mother Earth? When you buy menstrual products, you can see that there's some packaging involved. And menstrual products themselves are made up of paper and cotton materials. Keep in mind that products are made of natural cotton fibers, which can biodegrade (break down) in landfills. This is definitely a good thing when it comes to the environment. You might want to look for tampon products with biodegradable applicators. These products generally are labeled on the front of the package.

If you want to be extra sure that you're being friendly to the environment, here are some steps you can take:

- ❀ Dispose of your products properly!
- ❀ Recycle the cardboard boxes from your tampons or pads.
- ❀ Avoid flushing your products down the toilet—not only can this clog the plumbing, but improperly disposed items just might end up in the ocean!

Speaking of Supplies . . .

No one needs to tell you that it usually pays to be prepared—especially when it comes to your period. Sure, you can go to the school nurse, or maybe you'll be able to buy a pad in the school restroom vending machine or ask a friend if she has a pad to spare. But sometimes you might find that these plans can fail. It's best to be responsible for yourself by thinking ahead.

Planning Ahead

To avoid any menstrual mishaps, plan on stashing supplies in a few different places so that you'll have 'em on hand when the time comes:

- ✿ Bathroom cabinet
- ✿ School locker
- ✿ Gym locker
- ✿ Backpack
- ✿ Purse

Optional places include:

- ✿ Sports bag
- ✿ Glove compartment of your mom's car
- ✿ Makeup bag

Super Storage Ideas

Do you plan to store supplies in areas where others might see them, such as

. . . a shared locker?

. . . a bathroom used by other members of your family?

. . . the glove compartment of the family car?

. . . your purse, which might accidentally spill open at some point?

If so, you might want to place your supplies in a carrying case of some sort. Some girls like to use small hard-shell plastic cases or little makeup-style bags with zippers. People shouldn't be rummaging through your personal stuff—but things happen. Like maybe your best bud goes into your purse looking for something, or your little sister rummages through the bathroom closet. If this happens, just shrug it off. But having a case or a bag to stash your stuff in can give you some additional peace of mind.

As you can see, part of managing your period involves selecting the products that work for you and putting them in places where they're handy. Once you figure out the what's and where's, you'll be on your way to keeping your cool when it comes time for another menstrual cycle.

Fresh Factors: Keeping Clean

You probably already have figured out that you want to feel fresh and as clean as possible at all times. And during your period, this can be hard to do, especially on heavy flow days.

When it comes to feeling fresh, it definitely makes sense to change pads and tampons frequently. So how often is often enough? Changing your pad or tampon every three to four hours will help keep odors and leakage from happening.

Tampons are a little trickier than pads, so be sure to change them at least every few hours. While you can tell by looking whether a pad has soaked up all it's going to soak up, you can't check to see if a tampon is ready to be replaced. For this reason, you need to keep track of how long it's been since you put it in. This will help you avoid a nasty surprise like a leak. Also, leaving a tampon for a long time might result in a serious infection called Toxic Shock Syndrome (read more about this in Chapter 10).

Helpful Hints

- ❀ Put on a fresh pad before you go to bed. Replace it in the morning as soon as you wake up.
- ❀ Put on fresh protection (pad or tampon) before you go somewhere where it's a hassle to change. For example, maybe you're going to the beach or outdoor space where bathrooms

are hard to come by. If so, be sure to give yourself maximum time by replacing your protection just before you head out of the house.

❀ Dispose of applicators, outer packaging, and used tampons and pads by wrapping them in toilet tissue and throwing them out. In your home, you will probably throw them in a wastebasket. You might want to place them in a Ziploc-type plastic storage bag before tossing them into the trash. This can help cut down on odor.

Whatever you do, never flush pads down the toilet! They can cause serious plumbing problems. Certain types of tampons can be flushed, but make sure the package is marked "flushable." If you're in a public restroom, look for a covered bin that might be specially marked for feminine products.

If you're in the great outdoors, say, on a camping trip, plan on bringing along some of those zippered plastic bags. You can put used pads or tampons in these bags and then place them inside a suitcase or backpack temporarily. Then when you get to a place where you can dispose of the bag permanently, you can remove the bag and toss it.

Girl's-Eye View: Periods of the Past

Poor women of the past! For centuries, menstruating women were told they were unclean. In some places of worship, they were forbidden to participate in certain rituals during their monthly periods. Some were told to perform certain cleansing rituals before being able to step back inside their places of worship.

Just How Clean?

You might have the feeling that during your period you're just not as clean as at other times of the month. That's simply not true. Unlike people of the past who believed otherwise, today people know that menstrual blood isn't "dirty" or poisonous. It does have an odor, which is perfectly natural and won't be very noticeable if you change you pad or tampon often enough.

That said, you will probably feel better if you take a little extra care during this time of the month to keep odors down and feel fresh. If you practice generally good hygiene, you don't have to fret about whether you're fresh enough. For this reason, you might want to bathe more carefully on the days of your period, making a point to wash well between your legs. If you've been wearing a pad and don't like the idea of blood in the water around you when you bathe, you can shower instead. Make sure you always wear clean underpants as well.

As you're shopping, you'll probably notice that there are several vaginal deodorant products on the shelves. They're available in different fragrances and come in all kinds of feminine-looking packaging. So what are these products and are they right for you? If you're healthy and take care of yourself properly, you really won't need extra products like deodorants to help you feel fresh.

To Douche or Not to Douche?

Douches are products designed to cleanse the vagina from within. While you might be tempted to try them, keep in mind that the vagina is self-cleansing. Your body works hard to maintain a chemical balance to help reduce the chances of infection. In the event that you do get an infection, a doctor might tell you to use a douche for medication purposes. But in most cases, you don't really need to buy over-the-counter douches.

Change your protection when necessary. This helps you feel fresh and keeps odors in check. If you've experienced some leakage, like blood on your underwear or maybe on your inner thighs, this leakage can come in contact with bacteria and increase the odor. So if this happens, you'll probably feel tons better by taking an extra shower. The main point to remember, no matter what, is this: Don't get caught up in the mistaken idea that you're yucky and dirty. Just take extra steps to baby your body a little bit and feel fresh so you can feel good all over!

Menstrual Myths Through the Ages

Girls today know that it's important to be informed when it comes to the subject of menstruation. They're lucky enough to have scientific evidence to back up this body talk, so they can make smart decisions. Most of the real girls on this panel learned the real deal about menstruation early. By the time they heard myths about menstruation, they were able to tell right away that these were simply tall tales that were anything but truthful.

" "Sure, I heard some stories, but I knew better than to believe them," says Ali. "In fact, I can't even remember any of them. That's how much attention I paid to them. "

But consider the girls who came way before you. They weren't so fortunate. Science hadn't yet provided accurate information at the time they were growing up. Girls in your grandma's day and way before didn't have the facts; they just had guesses and maybe even falsehoods. And yikes! Check out the untruths they had to deal with!

MYTH: Ancient Romans believed that female horses (mares) miscarried if a menstruating woman touched them.

MYTH: A famous writer of ancient Rome wrote that bees left their hives if a menstruating woman touched them.

MYTH: Hippocrates, an ancient Greek physician, told his people that women's periods served to rid women of bad moods.

MYTH: In many cultures, people were taught that menstruating women could cause crops to fail and wine to turn into vinegar.

MYTH: Women who were menstruating were thought to cause milk to go bad and food to spoil in some cultures.

MYTH: Some cultures believed that clocks stopped because of menstruating women.

MYTH: In Australia, aborigine boys were told that if they saw menstrual blood, their hair would turn gray.

Maybe you've heard a few myths passed around at your school. If someone tries to tell you something that seems to be more fake than fact, be sure to check it out with a trusted adult and get the truth. No need to walk around freaking yourself out by believing everything you hear!

9

The Lowdown on PMS: What to Do and How to Deal

It's easy to confuse the terms "menstrual period" with "menstrual cycle." But there actually is a distinction between the two. The menstrual cycle involves the action of the hormones and the shifts in your reproductive organs during the entire month or so, including the period of time in which your menstrual fluid flows. A menstrual period, on the other hand, describes the actual days when you're discharging menstrual blood.

Before my period, I get this huge craving for junk food. I don't know why because I don't normally eat much junk food. But when my period comes, forget it. I've gotta have something really junky. I get bloated as well, irritable and crabby. It's just not fun to be me during those times. Sometimes I just don't want to be around ANYONE. — Trish

Premenstrual Preview

The days before your period are referred to as the premenstrual days. You may experience premenstrual symptoms—that is, a few days before your period, you'll notice signs that your period is about to begin.

On the one hand, you may find that your energy levels go up during the days leading up to your period. You might experience a sense of wellbeing and have the feeling that everything is going your way.

On the other hand, you may find that the symptoms you experience before your period aren't so pleasant. Check out the list below for examples of symptoms you might have. You might find that you don't experience any of these symptoms, or maybe you just have one or two. Sometimes the symptoms you encounter cross the line from things you hardly notice to things that really make you uncomfortable and cause you real concern. This is called Premenstrual Syndrome, or PMS.

Mood Swings

Maybe you're the type of girl who considers herself to be pretty levelheaded. When everyone else around you is flapping, you remain unruffled. You generally can keep your cool when things around you get crazy. When people bug you, you shrug it off, no problemo. When something bad happens, you feel sad, but you find it easy enough to move on.

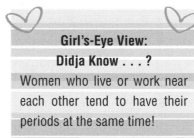

Girl's-Eye View:
Didja Know . . . ?
Women who live or work near each other tend to have their periods at the same time!

Yet, right before your period, all that changes in a big way. One minute you're tight with your best buds and feel like you're so lucky you have such great friends. The next minute—for no particular reason—you feel like they're your worst enemies! Or maybe one minute you're feeling lighthearted and ready to take on your day. The next, you find yourself feeling down and wanting to crawl under your covers and stay there all day instead! No . . . you're not going crazy! These feelings are called mood swings, and they're a perfectly normal part of PMS.

Backaches

Ugh. Nothing like a dull, throbbing ache in your lower back area to make you feel blah. It's not exactly painful, but your back's definitely letting you know in a big way that it's there.

Bloating

Just before your period, you may feel yourself swelling up like a blowfish. Your normally flat tummy seems to protrude, and it seems kind of sensitive and puffy. Putting on your jeans becomes a bit of a production. Seems that you can't zip 'em up as easily as you usually do. What can you do? Blame it on bloating.

Swollen Breasts

Your breasts seem to be a little bigger than usual. They're tender to the touch. Sometimes they downright hurt. No doubt about it, during PMS, your breasts aren't the same old breasts you've gotten used to. The days leading up to your period, it's like they're announcing their presence in a way you'd rather they didn't. Maybe some days, you wince at the thought of putting on a bra.

Headaches

Talk about a pain in the head. You're going about your day, and you feel like your brain is banging against your skull as you walk. Maybe it's not like *pain* pain, but it's definitely got your attention.

Fatigue

Yawn. A few days before your period, you might notice that getting out of bed in the morning is a big chore. And as you go about your day, you might find yourself feeling super-sleepy and lethargic. Every little task might feel like a big effort. You just want to plop your head on your desk, close your eyes, and float off to snoozeville.

For years, girls were told that premenstrual symptoms weren't real. They were told that these symptoms were "all in their head." Today, health-care professionals recognize that there actually are real premenstrual symptoms.

A very small percentage of girls experience symptoms so severe that they are unable to function normally. Generally, most girls find that their symptoms don't cause them serious concern. Note: If your symptoms get in the way of your usual routine, it's very important that you don't try to ignore them and hope they'll go away. You need to bring your concerns to your mom's attention so that she can make an appointment with a health-care professional.

Got Premenstrual Symptoms?

Bloating. Face Freak-outs. Backaches. Headaches. Mood Swing City. Think just because you suffer with these symptoms you've gotta put up and shut up? No way! You can fight back and make it so these bummers don't get the best of you. Here's how!

Educate Yourself

Find out what things trigger your symptoms so you can make some changes to avoid them. To do this, write notes or keep a calendar about what your exercise, sleep, and eating habits are on the days leading up to your symptoms. You might start seeing a pattern develop. For example, you might notice that you don't get enough sleep on certain days, and that a lack of sleep makes you more prone to feeling down or lowers your tolerance for pain.

Menstrual Myths

Okay, you're an informed, growing girl, and you're not going to let silly myths stand in your way, right? Right! Knowledge is always the best defense, so keep yourself informed by de-mything even more myths about menstruation.

MYTH: You're not clean during your period.

FACT: Menstrual blood isn't dirty. You're just as clean as when you're not menstruating.

MYTH: You can't exercise during your period.

FACT: There's no reason you can't exercise during your period. In fact, exercise might help you feel better.

MYTH: All girls feel tired during their periods.

FACT: Lots of girls experience no energy drop during their periods.

MYTH: You can't swim while having your period.

FACT: If you use tampons, there's no reason to stay out of the water while you're menstruating.

Girl Talk

Talk with your friends about what symptoms they might experience. They might have some remedies that you might not have thought of. Share the things you've learned as well. It might surprise you to find that lots of girls have come up with different ways to feel better. Some of them might be useful for you.

Change Your Habits

Once you've figured out which things make your symptoms worse, take action! You can sidestep those symptoms by making a few changes.

Exercise

You might feel like flopping on your bed and not moving a muscle during your period. And forget gym class, you say? The last thing you may feel like doing is running laps! When it comes to jumping jacks, count me out, you might mutter. Wait a minute. Being a slug might not be your best bet when it comes to feeling better.

Here's the deal: Lots of girls say they feel much better when they get up and start moving. You don't have to go all-out and try to run a 5K or engage in strenuous exercise if you can't muster the energy. You can just do simple stretches. If you find yourself starting to rev up, go for a slow jog. You might find that when your blood gets pumping and your head starts clearing, you'll feel recharged and ready to take on anything.

Get Your Zzzzzs

There's no doubt that getting a good night's sleep makes sense any day of the month, whether you're menstruating or not. If you shortchange yourself in the shuteye department, you'll find that you feel irritable even at the best of times, and menstrual blahs will seem bigger than ever. So if you find you skimp on sleep during the days before your period, make sure you hit the sack earlier than you normally would so you can catch up on needed zzzzz time. It could help you stay sunnier than you might otherwise. It also may help you put up with any pain you might have.

Food and Your Moods

You are what you eat. That's especially true when it comes to your menstrual cycle. Read on to find out about foods that are good and bad to eat throughout your cycle.

Say No to Salt

If you tend to load up on salty foods during the days before your period kicks in, you might notice that you look and feel puffier than ever. And if you keep right on sprinkling on the salt during your period, you'll keep right on battling to button up your jeans. This is because eating salty foods causes your bodily tissues to retain water. That means you'll end up bloating a bit more than you might otherwise. Even worse, you also might find that salt contributes to feelings of tension.

You can increase your comfort level by avoiding these foods at period time:

❀ Salty popcorn
❀ Peanuts
❀ Potato chips
❀ French fries

Also, set down the salt shaker at the dinner table—you'll be amazed at how much it helps!

The Buzz on Caffeine

What's another culprit that can cause swelling and pump up your bloat factor? Caffeine. That's right. Coffee, certain teas, colas, and chocolate are loaded with this chemical. So go herbal for a few days. Choose herbal teas or fruit juices over soda or coffee. Your body will thank you for it!

The Real Deal on Junk Food

When it comes to face freakouts, you might find that you experience more flare-ups just before your period. Your skin might break out in a big way and seem oilier than ever. A common misconception is that this is caused by eating too much junk food.

Not so long ago, teens were told to lay off the chocolate and greasy foods in order to control these flare-ups. However, experts now agree that junk food doesn't cause skin skirmishes at all. This doesn't mean you should pig out before your period, though! Remember— caffeine, salt, and sweets can trigger other uncomfortable period symptoms. By eating more healthfully around period time, you will make your skin glow.

Sugar Shock

If you feel fatigued before or during your period, you might want to cut down on sugary foods and drinks. Sure, a sweet thing here and there won't harm you. It's when you go overboard in an attempt to battle the blues that you find out about sugar's shady side.

At first, sweets can fool you. The sugar they contain can give you a temporary energy surge. For a little while, you feel revved up and ready to take on the world. You might feel like shoveling in the sugary goods so that you can keep up the good feeling. Not a good plan!

Wham! Before you know it, your body processes this sugar. Then you crash and feel more lethargic and lazy than ever. So, think before you chow down on some of these snacks before or during your period:

- ❀ Sodas
- ❀ Candy
- ❀ Cookies

Whether you're menstruating or not, it's always your best bet to eat sensibly and healthfully.

Girl Gab: Real Tips for Beating the Body Blahs

Maybe you don't have particularly nasty menstrual moments. You count yourself lucky that you generally avoid bloating, cramps, and headaches. Yet you find that you have those blah feelings when you're having your period. Do you just have to put up with 'em?

Nu-uh. Check out these feel-better tips from girls who've been in Blahsville—just like you.

Cool Down/Pump Up

Sometimes you need a catnap; sometimes you need a pick-me-up. Jenny uses both when she's trying to beat the period blahs.

"When I start to feel myself getting emotional, I make a point to try to get a little extra sleep. I usually find that getting to bed a little earlier than usual or taking a quick nap in the afternoon will help me be calmer about everything.

At other times, if I feel fatigued, I might turn on rowdy music and sing aloud and dance. I find that moving around — even if I don't feel like it at first — pumps me up again so I can get on with whatever I need to do, like homework."

Blissful Bath

There aren't many woes a good bath can't cure—Kacie knows that!

"Since I get a few symptoms, I do different things to deal with them. I'll take long, hot baths if I feel bloated."

'Tude Tamers

When bad moods get the best of Ali, she takes a few steps back.

I know I shouldn't be around my friends when I start feeling my moods start to swing. I try to just kind of stay away from them. I'll eat at the end of the cafeteria table instead of being in the middle so I don't have to talk that much. It's kind of an effort to keep from snapping at people. I have to really bite my tongue at times. Sometimes I find myself clenching my fists in order not to say something snippy that I'll regret later.

As soon as I get home after school, I go up to my room and watch TV or a funny movie to escape. It makes me feel better to be by myself at those times.

Going Solo

Sometimes a few minutes alone can drastically improve your outlook, as Crissy tells us.

"I like to be by myself when I get the menstrual blahs. I'll go in my room and blast my headphones."

Cramp Conundrum

Sometimes cramps can really get the best of you! Check out Katherine's plan for cramp relief.

"Sometimes I have really bad menstrual blahs. I cramp a lot, especially when I'm ovulating on one particular side. I also get these pounding headaches. When this happens, I'll have my mom give me an over-the-counter medicine. Then I climb in bed with a hot- water bottle. I know it's kind of extreme, but I really do have some bad days. When I do, my mom understands and she lets me stay home from school."

Battling the "Blahs"

When Alex feels a bad case of the blahs coming on, she does things to help her relax.

"Whenever I start feeling like I'm not really myself, I drink hot tea with a little bit of sugar in it. I find it really soothes me and makes me feel better. I also like to play with my dog. Pets can really make you feel happier at moments like these."

Refreshing Remedies

If you're looking for even more options to beat the blahs, here are some other tactics you can try:

✿ Curling up with a hot water bottle or an electric blanket
✿ Aromatherapy
✿ Chatting with a friend
✿ Playing a soothing musical instrument

===== Period Pop Quiz =====

When it comes to dealing with unexpected menstrual moments, do you measure up or mess up? Check out these possible period scenarios and decide what you would do.

1. You're feeling a little uneasy about the idea of starting your period for the first time, so you:

    A. Cram the brochure about menstruation that you got in health class into your backpack. Sure, you're a little concerned about your period, but the brochure will still be there whenever your period arrives. When the time comes, you'll simply dig it out and read up on what you need to do next.

    B. Put it out of your mind completely and pray it doesn't happen for a while. Any time anyone like your mom or best friend brings up the subject, you try to start talking about something else.

    C. Read all you can on the subject, even if the day seems kinda far away. Anyway, you want to know ahead of time what to expect. You practice putting on a pad just to know what it feels like. After all, it's better to plan so you feel like you're in control when your period puts in an appearance.

2. You're walking through the mall with your best friend. You're in a hurry to hit your favorite store because the mall's about to close. All of the sudden your BF gets a look of horror on her face. "I think I've started my period," she whispers. "I don't have any supplies." You:

    A. Point out the restroom and let her figure it out herself. After all, it's not your fault she doesn't have a pad!

    B. Tell her not to worry. You always carry an extra pad in your bag in case of emergencies, and you're always willing to help a friend in need.

    C. Freak. She's just going to have to wait to deal with it till after you've shopped a little more—after all, the mall is closing in a few minutes, and she's cutting into your precious shopping time. Maybe afterwards, you'll go with her to the restroom to see if there is a vending machine. Anyway, she should have been better prepared.

3. It's late morning on the second day of your period, and it's been a few hours since you changed your pad. You have a short break. You could either kill time and hang out with your friends by the library or run to the restroom and freshen up by replacing your menstrual protection. You:

A. Really want to chill with your buds, so you figure you can take a chance that your pad won't be soaked by lunchtime. If it starts leaking you could probably avoid a real embarrassing moment by bolting for the restroom when the lunch bell rings.

B. Think for a sec about your menstrual history. You know your body, and you think you can make it easily till lunchtime without an accident. Then you head over to the library area to find your friends.

C. Decide it's better to be safe than sorry. Besides, you're wearing your fave jeans and there's no way you're going to take a chance on leakage. Anyway, you know that it's smart to cut down on odor-causing bacterial buildup by changing your pads or tampons frequently.

4. You're on a school field trip to a museum, and just as you've stepped off the bus, you feel wetness in your underpants. You:

A. Ignore this sensation and keep walking with your classmates through the rest of the exhibition. It would be too embarrassing to ask a teacher for permission to use the restroom, and you wouldn't want anyone to figure out what's going on.

B. Wait for a break and then hit the restroom for a double-check.

C. No way you're spending the next few hours uncomfortable and worried! Better to risk asking for permission to go to the restroom than risk having an accident. Chances are that no one will even notice anyway.

5. You're in a public restroom, and you need to toss some used menstrual protection. You don't see a bin within the stall, so you:

   A. Flush it down the toilet. So what if it clogs the plumbing? It's not your house, and you'll be outta there in a flash. No one has to know it was you.

   B. You wrap it up in layers of toilet tissue and leave it on top of the toilet tank. The cleaning person will clean it up, you figure.

   C. You wrap it up in layers of toilet tissue, step out of the stall, and toss it in the wastebasket near the sinks. No sense leaving your mess for someone else.

6. You tend to have periods with heavy flow and some uncomfortable symptoms. Your family is planning the annual summer vacation. You all have decided on a wilderness trip and will be backpacking through remote areas of your favorite national park. Your mom and dad ask you for your vote when it comes to selecting a date, so you:

   A. Say "I hate making decisions. Whenever you want to go is just fine with me."

   B. Try to remember when your last period was and mention a week that you guess might fall after your next predicted period.

   C. Check your menstrual calendar or diary and give your folks a date that you can be pretty certain won't fall during the time you have your period. Who needs the extra hassle while you're out in the wilderness?

7. You're a shy, quiet type of girl. The idea of talking with your mom or favorite aunt about something as personal as a period terrifies you. Yet you have some questions you're dying to ask. You just know it'll be an embarrassing and difficult conversation, so you:

   A. Make a promise to yourself to bring up the subject very soon with your mom or aunt.

   B. Put off this question-and-answer session. The idea of

talking it out just gives you the willies. You'd rather walk around with a bunch of questions swirling in your brain than to risk a totally awkward conversation.

C. Get your guts up and go for it. It's time, you decide firmly, to tackle the tough questions and free your mind from uncertainty! You take the very next opportunity to kick off the conversation.

8. You're an active kind of girl who loves to swim and participate in water sports. You'd really like to try wearing tampons, but you're afraid, so you:

A. Talk with your mom about it, get a box of tampons, carefully read the instructions, and take some time to practice inserting a tampon.

B. Give in to your fears and forget about it. Decide you'll simply have to avoid swimming on the days you're having your period because it's just too scary to consider trying tampons.

C. Talk with your mom, buy a package of junior or slender tampons, and read the manufacturer's instructions thoroughly. You're still feeling a little unsure, so you take the time to think about it a little more.

9. Your classmates still gossip about someone who years ago had a really bad menstrual mishap at school. The idea of having an accident like this at school freaks you out. You:

A. Keep a sweater tied around your waist at all times to avoid playing the starring role in the next campus horror story.

B. Worry that you'll have an accident but keep the worries inside, hoping that nothing like that ever happens to you.

C. Try not to freak yourself too much—after all, every girl has her own period fears. You figure you can cut down on the odds of an accident by being prepared. You're keeping your period calendar, so you have a heads-up on when you can expect the next one. You also keep a stash of fresh supplies wherever you think they'll be handy.

10. Ugh. You've got a strange vaginal discharge that's starting to irritate you. In fact, it's kind of itchy. You're kind of embarrassed to say anything to anyone, and you aren't sure what to do. You:

A. Immediately tell your mom so you can see a health-care professional. You know better than to ignore a possible infection.

B. Ignore it and hope it'll go away.

C. Treat the itch with an over-the-counter product and tell your mom about it right away.

Now it's time to find out how you did. The girl panel has rated the answers so you know what the score is. Add up your points and see which category you fall under.

1. A) 2 B) 1 C) 3   6. A) 1 B) 2 C) 3
2. A) 2 B) 3 C) 1   7. A) 2 B) 1 C) 3
3. A) 1 B) 2 C) 3   8. A) 2 B) 1 C) 2
4. A) 1 B) 2 C) 3   9. A) 1 B) 2 C) 3
5. A) 1 B) 2 C) 3   10. A) 3 B) 1 C) 2

If you scored 18 or fewer points: RED ALERT!

You're headed into the danger zone when it comes to menstruation. By turning your back on menstrual issues, you risk running into needless worry. Keep your mind open, and be sure to look and listen when it comes to dealing with periods.

If you scored 19–24 points: YELLOW CITY

You're on the right track but some of your signals are crossed. You might be a little hesitant to seek the answers you need. Make it a point to learn more about this important health issue and be ready to assist your friends when they need period-related help.

If you scored 25–29 points: GREEN-LIGHT GIRL

You're on the fast track, always ready to absorb information and pass it on when needed. Good for you. Keep asking questions and seeking answers. You'll be in better control of your health and able to lend a hand to pals when necessary as well!

10

Problem Periods: Quick Fixes for Every Situation

When Periods Are Problematic

Chances are, you'll find that having your period is no big deal. But if you do find that you experience problem periods, you don't have to freak. Read on for help in dealing with more-than-annoying menstrual moments!

Irregularity

Generally speaking, most girls like to know when to expect their periods so they can plan and prepare. If you never know when it's going to start, it can be a real annoyance. It might make you feel uneasy and less willing to plan activities. You might feel like you just shouldn't chance doing certain things, like planning that trip to the beach, because your period might pop in at any time.

Thing is, even if your period puts in an appearance when you least want it, you don't have to put a stop to your activities. If you bring along protection, you'll be ready.

Even though many girls experience irregular periods when they first start menstruating, you're probably wondering when your cycle will settle down and become predictable.

Many girls start to see their periods fall into routine patterns after a year or two. Still, even after that, it's quite possible that you will find some irregularity occurring. You might find that your period is unexpectedly delayed. At times, your period will seem heavier or lighter. It might be accompanied by several symptoms one month, and none at all the next.

Girl's-Eye View: Regular or Irregular?

Remember, even though periods may be irregular for the first one to three years, they should still be twenty-one to forty-five days. If you're skipping periods or getting more frequent periods, you should talk about it with your mom and your doctor.

Lots of things can affect your menstrual cycle and keep it from coming regularly each month. Some of these things include:

- ✿ **Stress.** If you're feeling stressed at school or home, for example, you might find that your period can be delayed.
- ✿ **Emotional ups and downs.** If something happens to upset you greatly, you might find that your period puts in an appearance earlier than you expected. Then again, it might not occur till days after you expected it.
- ✿ **Sickness.** Illness can play a part when it comes to menstrual irregularity. If you've had the flu or been under the weather, it might affect your period.
- ✿ **Physical activity.** Are you athletic? If so, you might find that playing some serious sports delays your period.
- ✿ **Traveling.** There's nothing like a trip away from home to throw your body a bit off balance. Your period might happen earlier or later. For this reason, you'll want to be prepared!
- ✿ **Weight loss.** If you lose a lot of weight, this can affect your periods. If you lose a large percentage of your body weight, you may even find that your periods will stop altogether. If this happens, it's important to see your doctor.

While most girls prefer that their periods be regular, it doesn't always work that way. Some girls never are able to pinpoint exactly when to expect their periods. For some girls, being irregular is just part of who they are.

Other Than Blood

You might find that on the days of heavy menstrual flow you see more than just blood on your pad. Sometimes along with blood, dead cells from the uterus are discharged as well. This discharge might look like little blobs of Jell-O. When you first see these blobs, it might make you nervous. After all, it can really look like a lot of menstrual material. You might wonder if something is seriously wrong. But keep in mind that there's nothing wrong. It's natural.

Cramps City

The good news is that lots of girls have cramp-free periods. Except for having to think about changing their protection regularly, they find that they don't have to think about menstruating at all.

The bad news is that some girls do experience cramping. They might not have cramps during every period, but many girls experience them at least some of the time.

Generally, most girls get cramps for the first few days of their period and find that they gradually diminish each day. This cramping is caused by muscles contracting. You might feel a tightening of the muscles in your lower abdomen area. This feeling might extend partway down your inner thighs as well. It might be a slight sensation or it might be strong enough to really get your attention. Some girls get severe cramps that require doctor-prescribed medications.

Girl Gab: Cramp Talk

Cramps are noooo fun, but many girls get them at one time or another. Here's what the girl panel had to say about their cramps.

Time Heals All Cramps

Emma learned that some cramps are only temporary.

> When I first started having periods, I got cramps. They were pretty strong and I found that they were really uncomfortable. Luckily, over time, I stopped getting them. They haven't returned in quite some time now.

Lucky Lady

We all wish we had Becca's luck!

> *I guess I'm just extra lucky. No cramps here. I never had them.*

Cramping Your Style

Depending on your body, cramps can be mild or very painful. Don't despair, though. As you get to know your body, you'll find tricks to beat your cramps, just like Jenny did:

"Ugh. Cramps! They can be so annoying. I feel them in my lower abdomen and sometimes the feeling goes down into my inner thighs."

Cramp Control

Most girls who get cramps find that they're merely a slight annoyance. You might discover that by staying busy, you'll soon forget about them. But what if you can't forget about them? Well, then it's time to come up with a battle plan. You might want to try a few of these techniques to find out which ones work best for you.

Out with the Old, in with the New

In your grandmother's day, girls who were menstruating often stayed home and avoided activity. Today, many girls find that their cramps subside if they exercise lightly.

Breathe Deeply

You might find that you can help control cramps by relaxing and using deep-breathing techniques. When the pains come, breathe in and out, slowly and rhythmically, till they let up. This might be hard to do because when you're hurting, your natural tendency is to tense up. Just relax and remind yourself that though cramps are anything but fun, they're simply the result of your uterus contracting.

At Your Fingertips

You might find relief by massaging your abdominal area. You can gently place your fingertips on the areas where you experience cramping. Move your fingertips in a slow, soothing, circular motion. You can increase the pressure or decrease it, depending upon what you find to be comfortable.

Turn Up the Heat

Put a crimp in your cramps by placing a hot-water bottle or a heating pad on your abdomen. Or, you might want to take a soothing, warm bath instead. Sipping warm herbal tea also can help.

Pain Pointers

Keep in mind that if you experience more intense cramps, headaches, or other discomfort, you don't have to simply grin and bear it. Pain does not have to be part of your period. Even if someone tries to tell you that "it's all part of being a woman," don't listen. It's okay to take steps to alleviate discomforts associated with your period.

If you take a trip to your local drugstore, you'll see that there are a number of over-the-counter medications that can help take the ouch out of your menstruating moments.

If these products don't seem to help and your headaches or cramps continue to cause you severe discomfort, see your doctor or health-care provider. These people can evaluate your needs and order various prescription treatments to help put a stop to your pain.

Girl's-Eye View: Period Prescription

Some newer pain treatments involve the use of "anti-prostaglandins." Big word, right? Anti-prostaglandins inhibit the actions of some of your havoc-wreaking hormones. They can help you become pain-free so that you feel healthy and can stay active no matter what time of the month it is.

Pre-Pain Strategy

According to Elaine Pan, M.D., a gynecologist who often sees young patients, girls who experience painful symptoms should talk with their moms and doctors about managing this pain. "It's best to be on top of pain, to deal with it before it starts," she says. "For example, if you're prone to cramping every month, you might want to take an over-the-counter medication the day before you start your period."

Infection

Even when you're not menstruating, your vagina can shed dead cells and mucus. You'll see signs of this discharge in your underpants. It will have a light yellow appearance and next to no odor. This is perfectly normal.

If you notice that the discharge takes on a greenish or brownish color or appears to be thick and lumpy, that's another matter. If the discharge gives off an unpleasant odor, that's also cause for concern. Sometimes the discharge might make you itch or even burn. You might notice a burning feeling when you urinate. Signs like this can mean that you have an infection. Should you have signs of infection like these, go see your pediatrician or a gynecologist right away.

Girl's-Eye View: Protect Yourself

You should never let an infection go untreated. Infections that go unchecked can get out of control and really do a number on your health. So don't be shy about telling your mom what's up right away so she can make an appointment for you to get you the help you need.

Toxic Shock Syndrome (TSS)

Toxic shock syndrome is a rare but extremely serious condition that tends to come on suddenly in a dramatic way. A large number of women affected by this condition are under the age of nineteen. This condition first came to national attention when a number of menstruating women were affected. Symptoms can include high fever and vomiting.

So what causes it? Experts are still trying to understand and pinpoint the exact causes of toxic shock syndrome. It's believed to be caused by toxin-producing bacteria called *Staphylococcus aureus*. The body responds in a dramatic way to the toxin and goes into what's called "hypotensive shock." This means the heart and lungs stop working. Scientists believe that there is a connection between toxic shock syndrome and super-absorbent tampon use. They studied particular menstrual products and pulled some off the market that they believed contributed to cases of toxic shock syndrome.

If you have further questions about toxic shock syndrome, talk with your health-care professional.

Swollen or Sensitive Breasts

Though you've probably noticed that breasts are always pretty tender to the touch, you might also notice that they are especially sensitive just before your period starts. They might appear to be slightly swollen. Sometimes, they're downright achy or even painful.

If you're bugged by breasts like these, take extra care during this time of the month. If you find that your bra irritates you, you can take it off for a few days. If, on the other hand, you find that bouncing around bothers your breasts even more, try wearing a bra with more support on these days.

Headaches

Those mild headaches you might experience occasionally might not be random. If you write down on your calendar that you get headaches, you might see that they occur while you're menstruating.

"I get nasty headaches during my period," says Katherine. "They're the pounding kind that seem to be right in my temples."

If you're like Katherine and you're prone to menstrual-related headaches, check with your doctor. She might suggest that you take an over-the-counter pain reliever. If your headaches are more severe, your doctor might suggest additional treatments.

Heavy Periods

Some girls experience heavier periods than others just because of the way their bodies work. According to Dr. Pan, some girls who are obese find that their periods tend to be heavy.

If your periods are heavy and this causes you to worry, check this out with your doctor.

Girl's-Eye View:
Toxic Triggers
Scientists believe that toxic shock is triggered when bacteria enters the bloodstream, and that keeping a super-absorbent tampon in the vagina for a very long time can contribute to bacterial growth.

Extreme Moods

Happiness, sadness, anger. Over the years, you've probably experienced emotions like these. It's normal to feel a wave of happiness when something good happens, like when you get a good grade on a test or when someone does something nice for you. And it's perfectly natural to feel disappointed and a little sad when something doesn't go your way.

Some girls find that during the days leading up to their period, they experience an emotional roller coaster. You too might find that every month for a few days your emotions suddenly become supersized! Maybe you feel more sensitive and like every little thing is now a BIG deal.

Girl Gab: Period Pains

Could be that you, like some of the girls on our real-girl panel, have cause to complain a little just before your period arrives on the scene. Maybe you experience annoying symptoms like the following, which mean one just thing: your period is just around the corner!

A Little on Edge...
Rachel gets a little grouchy, even when she doesn't want to!

"I find that I get a little more irritable before my period starts. Stuff that wouldn't normally bother me really makes me edgy. I wish it wasn't that way because I don't like feeling grouchy, but it is."

Cry Me a River
Jenny gets a little teary-eyed around that time of the month.

"I always know when my period is coming. A few days before it starts, I start to cry at the stupidest things. For example, I'll sob my eyeballs out when I watch a TV commercial where they play sad music. I'll cry if I drop something or if it takes me an extra few seconds to find my volleyball sweatshirt or something. Even though I know it's just my hormones making me all weepy, it bugs me to feel this way."

Hiding Out
Sometimes, bad moods get the best of us, even when we try not to let them. When that happens, you might want to take a time-out, just like Trish.

"I find I get so crabby, I don't want to be around *anyone*. It's just too hard right then. That's when I go hide in my room."

Feeling Fatigued

Like Becca, you might feel a little tired just before your period.

"Before my period, I feel a little tired. During the first couple of days, I still feel kind of tired, but I still go play volleyball and stuff. I don't get cramps like some of my friends do."

As you can see from these girls' stories, your body is a little different at period time. Things that you might have taken in stride before now seem like life-or-death matters. Instead of feeling mildly disappointed when your soccer team loses a match, you feel like it's the worst thing ever! Instead of feeling a little down that your friends didn't think to ask you along on their mall trip, you feel like bursting into tears!

What's going on here? Why are your emotions taking you on such a wild ride?

It's actually pretty simple. When hormones fluctuate, your moods can fluctuate with them. When you're starting to go through puberty, your hormone levels go crazy. In most cases, girls find that their moods even out as they move into the later stages of adolescence.

Does It Ever Get Easier?

By now, you've gotten a pretty good idea of what to expect when it comes to the changes your body's going to go through. You know what causes the changes and what it'll mean to you. But maybe you've decided that when it gets right down to it, you still really just don't want your body to change.

That feeling is perfectly normal too. After all, you've had your little-girl body all your life and it's worked perfectly fine that way, thank you.

Quite a few girls find that they wish things would just stay the same. They might be thinking: "Who needs to worry about remembering to bring pads everywhere?" "Who wants to have to think about changing pads?" "Who wants to have to try to remember when to expect their next period?"

While it's true that starting your period means having to plan and prepare, it's also true that you'll get used to the process pretty quickly. Before you know it, you won't think twice about menstruating at all. It just becomes a normal part of your life.

Girl Gab: Been There, Done That

Getting your period for the first few times can seem like a real biggie. But after a while, you'll find yourself knowing what to expect. Soon having your period becomes almost automatic. You might find, like the real girls here, that sure, you take care of yourself during this time, but to think much more than that about periods? Naw . . .

A Normal Occurrence

For Alex, getting her period has become part of her routine.

It's kind of like brushing your teeth or shampooing your hair. I never think about it. It's just a natural part of my life. It's like breathing. Just something that happens while you're going about your usual routine.

Perfect Prep

Ali makes sure she's prepared for her period at all times.

"I don't think about it anymore. I just make sure to carry a lot of supplies with me so that I'm prepared. That keeps me from thinking about menstruating any more than I have to."

Keep On Movin'

Rachel doesn't let anything get her down—especially her period!

"Menstruating is something you worry about ahead of time, maybe. But then you just take it as it comes and try to keep it from ever slowing you down or getting in the way of your life."

Busy Buzz

Jenny's a busy girl, and she doesn't let her period get in the way.

> These days, getting my period every month is a total non-event. To me, it's just like, 'Oh, yawn, here it is again.' I have to bring along supplies or whatever and change from time to time. But beyond that, I totally forget that I'm having my period at that time of the month. I guess I'm just too busy living my life to stop to think about something that's just not a biggie.

You and the Gynecologist

You already know that most girls entering their teen years begin to have questions about their bodies. As you go along, you might begin to have specific concerns about things such as infections, the size and shape of your sexual organs, lumps in your breasts, or period symptoms. If so, you'll want to go see your family doctor or pediatrician. If you have health issues involving your period or issues involving your reproductive organs, your doctor most likely will refer you to a gynecologist. A gynecologist is a medical doctor, one who specializes in dealing with women's bodies and related health issues.

Girl's-Eye View: Going to the Gyno

A gynecologist often is an obstetrician (a doctor who deals with women before, during, and right after childbirth) as well as a fertility specialist. As a fertility specialist, she might help couples who are having difficulty conceiving (creating) a baby.

Why Should Girls See a Gynecologist?

Every girl should see her gynecologist once a year. This is so the doctor can check to make sure that all the sexual organs are functioning properly. Chances are that your mom has a gynecologist whom she sees regularly. You might feel comfortable in going to this gynecologist as well.

When Should I See a Gynecologist?

Generally, Dr. Pan says, girls should start seeing a gynecologist for regular exams at age eighteen or when they become sexually active, whichever comes first.

However, if you have any health concerns related to your reproductive organs, you should see a gynecologist right away. It doesn't matter how old you are. Sometimes even girls who are tiny toddlers are taken to a gynecologist if they have specific health issues related to their female organs, such as vaginal infections.

Before your first gynecological exam, you should find out what to expect. You don't need to whip yourself up into a worryfest by not knowing what goes on during an exam.

Start by talking it over with your mom. Have her describe a typical exam to you. Ask her any questions you might think of. "I find that girls who discuss this subject with their moms before their first exams tend to be more comfortable," says Dr. Pan. "Even the girls whose moms are in the room with them for the exam usually seem to find the procedure easier if they discussed what to expect beforehand."

What should you expect at your gynecological checkup? Though it might seem a little scary at first, rest assured that there's nothing scary about it at all.

Doctor Dialogue

Kathryn Iwata, M.D., a gynecologist who practices obstetrics and deals with infertility (which is when people have difficulty making a baby), says, "First of all, decide if you want your mother in the room with you—not a bad idea when the doctor is taking a medical history. [Maybe you'll want] to be alone when the doctor is doing the actual exam."

Girl's-Eye View: Hey—It's Okay

The good news is that the gynecologist won't think your questions are silly or embarrassing. After all, this is what he or she is trained to do. Though you might feel awkward if your gynecologist is a male, keep in mind that there's no need to be embarrassed.

When you first step into your gynecologist's office, she will ask you general questions about your health.

Dr. Iwata says that your doctor might ask you these questions:

- When did you start your periods?
- How often are the periods—monthly (regular) or whenever (irregular)?
- How many days do you bleed?
- Do you use pads or tampons?
- How many do you use when the flow is heavy?
- Do you get cramps with your periods?
- If you get cramps, what do you do to relieve them?
- Do you get PMS—irritable, anxious, depressed, bloated, acne, food cravings, nauseated, weight gain?
- Do you take any medicines for other medical problems (asthma, depression)?
- Are you allergic to any medicines?
- Have you had any surgeries?
- Have you been sexually active? If so, with the same partner, or different partners?
- Do you use any contraception?
- Have you been pregnant?
- Do you smoke, drink alcohol, or do illicit street drugs?
- Have you ever had a sexually transmitted disease?
- Do you have any family medical problems—parents, grandparents, siblings, with diabetes, high blood pressure, breast cancer?
- The doctor will probably also weigh you and measure your height.

Examination Situation

Next you'll get an exam. Your gynecologist will first do a general exam, just like a regular doctor. She'll check your heart, lungs, abdomen, blood pressure, and weight. Then she'll do a breast exam to check your breasts for uncommon bumps or lumps. Next comes the pelvic exam. It might feel a little, well, strange. That's because the pelvic area tends to be sensitive. Some girls might find an exam to be a little uncomfortable, but it won't hurt. So relax and know that being examined is just a perfectly normal part of being a woman.

When to See a Doctor

As you now know, there's a wide range of things that are considered "normal" when it comes to menstruating. So when should you be concerned enough to go to a doctor or other health-care provider?

Here are some signs you should not ignore:

- ✿ If you have been regular and miss more than a period or two.
- ✿ If you have lost a great deal of weight and your periods have stopped.
- ✿ If you are bleeding excessively.
- ✿ If your period keeps going for more days than usual.
- ✿ If you experience headaches that you believe might be related to your cycle.
- ✿ If you bleed more than a few drops between periods.
- ✿ If you experience cramps so painful that ordinary measures don't seem to help.
- ✿ If your period hasn't started by age sixteen.

As part of your pelvic exam, you will be asked to take off all your clothing, including your underwear. You will then be asked to sit on the examining table and lie down on your back. The doctor or nurse will place a light paper covering over you. Your feet will be placed in special pieces of equipment at the end of the examining table. These pieces of equipment are made of metal and are called "stirrups." You will be asked to scrunch forward so that your knees are up. This position, while it might feel weird or even kinda embarrassing, enables the doctor to examine you.

A good way to overcome nerves is to involve yourself in the process. Don't be afraid to ask questions. Have the doctor show you the size of your vaginal opening with a mirror. Have your doctor show you the piece of equipment called a "speculum." If you're a very young teen, the doctor will use a very small pediatric speculum.

This device is inserted into the vagina and is used to spread the walls so the gynecologist can examine this area.

Relax!

If you find that you're feeling nervous or embarrassed while the exam is going on, try to relax. Some techniques to try:

- ❀ Breathe normally—don't hold your breath or you'll be squeezing (which closes your vagina).
- ❀ Don't raise your hips or squeeze your buttocks.
- ❀ Keep in mind that there's lots of room in the vagina. "After all," says Dr. Iwata, "A newborn baby fits through there."
- ❀ Occupy your mind by playing a song in your head, or run through the storyline of a book you've just read.
- ❀ Picture yourself in your favorite relaxing place. This could be a vacation spot, like the beach or the mountains. Imagine yourself resting and soaking up the scenery.

Before you know it, your exam will be over!

The speculum will be inserted into your vagina. This doesn't hurt; it's just a strange sensation. "There are no sharp objects like needles," Dr. Iwata tells us. "It feels just like putting a tampon in."

Even if you're nervous, Dr. Iwata advises, "Don't hold on to your mother's hand during the exam. Keep your hands relaxed at your sides. Squeezing hands . . . causes you to close off your vagina."

The speculum will open up the vaginal area slightly, pushing back the walls. The doctor then uses a flashlight to look up the vagina and reaches in with a finger. He or she will press on your stomach and feel your uterus and ovaries. This is to check for anything that doesn't seem quite right. This is also done to feel for any unusual lumps that shouldn't be there.

Girl's-Eye View:
Pelvic Preview

Before your pelvic exam, have your doctor demonstrate how the speculum opens—"Like a duck's bill," says Dr. Iwata. This will help you feel more comfortable with the exam.

The doctor might also take a cotton swab and reach inside the vagina to take a routine tissue sample. This sample is taken to a lab for testing for possible infections. This procedure is called a "Pap smear." If the doctor does a Pap smear test, it's not necessarily because he or she thinks you have a problem. This is just part of the typical exam.

When your exam is over, the doctor will ask you to dress again. Once you're dressed, most doctors check again to make sure you don't have any further questions. If there's anything you'd like to ask, no matter how small or weird it seems, now's a good time to ask. You'll feel better being able to air your concerns with a doctor and seeing what the doctor has to say. It's nice to have an educated opinion when something's on your mind and to put any worries you might have to rest. Rachel's reaction:

"When I first had an exam, I admit I was hyper nervous. I felt kind of sick to my stomach, I was so keyed up. I was mumbling things to myself, like 'Get me outta here!' But then I decided it would definitely be better to stop saying that and to say other things to calm myself down. I'd say things like, "You're okay. It's all okay. You'll be fine.'"

11

Less-Than-Perfect Period Moments: Tell-All Tales!

For ages, girls have sat around campfires at camp or gathered around a flashlight at a sleepover and have whispered some outrageous stories having to do with menstruation. Most of them are made up, but some are based on truth.

It seems that everyone's got a menstrual story to tell. While most menstrual moments aren't that momentous, some can be downright mortifying! They're the moments when you feel like everyone is looking at you and your whole world has gone wacky. These are the times you're not sure you're going to live through when they're happening.

So why read about periods that made for some embarrassing personal headlines for other girls? First, so you'll realize that things can happen even though you wish they wouldn't. These are things that can usually be prevented with a little more preparation. These stories and the tips that follow will teach you how to take the extra steps you need to cut down on your chances of experiencing embarrassing menstrual moments.

Second, these stories will show you that you can live down embarrassments pretty quickly if you take the right attitude. Learn from these real girls. Sure, they had some less-than-perfect periods, but they shrugged it off and went on quite nicely with their lives—and so can you! So work that positive attitude, girl!

Friend Fix

"Once when we were in eighth grade, my friend was wearing light, khaki-colored pants. She stood up in front of me in class at school, and I could see blood all over the back of her. I felt bad for her. This was like my worst nightmare. I wanted to save her from the massive embarrassment of having everyone notice.

I didn't make a big deal over it. I simply stood up quietly and tied my dark sweatshirt around her waist. She didn't understand what I was doing at first, but she glanced down and figured it out right away. She gathered up her books when the bell rang, and went to the office to call her mom for a change of clothes. No one knew anything. My friend thanked me later for being supportive and helping her out." — Rachel

Problem-Prevention Strategy #1

If your periods are irregular, try to wear darker-colored clothing if possible on days you suspect your period might come.

Problem-Prevention Strategy #2

If your friend's in need, try to ease her worries by coming to her aid quickly. Be calm and help her out without drawing attention to her period plight.

Problem-Prevention Strategy #3

Be a real pal: You can help a friend in a period predicament by not drawing attention to her situation. Grab a sweatshirt or walk behind her or whatever it takes to quickly get her to a place where she can address her problem. Afterward, assure her that it could happen to anyone and that it's no big deal.

Close Call

"My friend was wearing her school uniform, which was light colored. She forgot to change her pad and when she looked down, she suddenly saw that she had blood on her clothes. She said she wanted to die when she realized that a boy was looking at her and that he'd seen she'd had

(continued . . .)

an accident. But the guy never said anything, so my friend felt better. At the time, she just covered it up quickly by holding her books and she dashed for the pay phone to call her mom." —Becca

Problem-Prevention Strategy #1

It's important not to forget when your pad or tampon needs to be changed. Be sure to visit the restroom when you can during the school day so you can change as needed.

Problem-Prevention Strategy #2

If you stain your clothes and you're wearing a skirt, try turning it around. This way the stain will be in front. You can then hold your books low in front of you while you head to the phone to call your mom for a change of clothes.

Problem-Prevention Strategy #3

If someone (even a guy) sees that you've had an accident, try not to let it ruin your day. Remind yourself that guys go through some of their own embarrassing moments. If he makes a crack, respond with a little humor and then move on. Chances are, he'll forget it right away. If not, keep in mind that creeps usually end up having humiliating moments themselves.

Brave Babe

"One time when I was in eighth grade, I'd forgotten that it was about time to start my period. In the morning, it had crossed my mind and so I'd put on a panty liner, but that was as much as I thought about it. I went about my day at school and my mom came to pick me up. As I was climbing into the car, I noticed that my pants were soaked. Oh, my gosh, it was gross! All I could think about was, 'How long had I been like this? All day? Did anyone notice? How come no one told me?'

That night, I kept thinking that I'd have to skip school. There was no way I could face people. But then I decided to brave it and say nothing. I figured if I said nothing, even if people saw anything, they might not say anything either, and the whole thing would be forgotten. I was right."
— Trish

Problem-Prevention Strategy #1

Be prepared. If you're menstruating, you don't have to completely change the way you dress. Remember, people can't tell you're wearing pads if you put them on correctly. But you might feel a little more confident avoiding form-fitting pants anyway. What's more, you might prefer to save white clothing and light colors for no-flow or lighter-flow days. You'll probably feel more comfortable wearing darker, looser-fitting clothing. It helps to know that if you experience leakage or have an accident, it won't be so out there in plain view for everyone to see.

Problem-Prevention Strategy #2

Think ahead: You can cut down on the risk of leakage by changing your protection more often. Check your stash of supplies each morning before you go to school. Carry pads, tampons, and panty liners with you in your backpack, purse, or sports bag so you have them whenever you need them.

Problem-Prevention Strategy #3

Keep your pal's predicament to yourself. No need to advertise the fact to anyone, even other friends. She'll be glad you respected her privacy, and you can feel better knowing that should the situation happen to you, she'll be more inclined to keep your "crisis" quiet just as you would want her to.

Problem-Prevention Strategy #4

Be a forward-thinking friend: It's great to be prepared yourself so that you can cut down on your chances of having an accident. But it doesn't hurt to be ready in case a buddy has an emergency. Why not throw an extra pad or two in your locker just in case you have a friend in need? She'll be grateful—that's for sure. And she might learn a little something from you so she can return the favor in case you're the one caught unprepared one day!

Beach Bummer

"Once when I was younger, I was with the family on a beach vacation. My period started unexpectedly, and I didn't have any pads with me. My parents were down on the beach with my little brother, so I was on my own. I had to ride my bike to this little store on the corner that sold sunglasses and sunscreen as well as toiletries.

Imagine how I felt when I walked in and the only person working at the store was this totally cute teen surfer guy. Worse yet, the menstrual supplies were behind the cash register. I wanted to crawl away and die. But there was no getting around it: I had to have some pads. I just looked the guy in the eye, acted like I didn't care (even though I was completely dying inside) and asked him for a box. He was cool, though. Still, I got outta there so fast!" — Jenny

Problem-Prevention Strategy #1

Remember, guys generally have to take health class, too, so they end up learning at least something about menstruation. What's more, lots of guys have sisters and they probably see their sisters' menstrual supplies in their bathrooms at home. So keep that in mind if you're in a situation like Jenny's and have to brave asking a boy when it comes time to purchase pads.

Problem-Prevention Strategy #2

If the guy's a creepo supremo and makes a rude remark or gives you a weird look, tell yourself that it just doesn't matter what he says or does. Menstruation is a natural thing. You didn't ask for it. It just is. And anyway, it's not like you oughta care about what a total jerk thinks anyway! Pu-leeze!

Problem-Prevention Strategy #3

Going on vacation? Even if you're somewhat sure that your period isn't due during the time you'll be gone, play it smart and pack some pads or tampons in your suitcase anyway. When you're on the open road or at a vacation spot, things just aren't as predictable as they are at home. If you're traveling somewhere far away, say a foreign country, for example, you might have trouble locating supplies. Or you might not realize it's a holiday and that the local stores will all be closed.

Even if you're staying in your own country and simply traveling somewhere that's close by, you can't be sure the local stores will be open when you need them. Vacation spots and small towns don't always operate on the same timetables as bigger cities. You can't always be sure that local stores stock the menstrual products you prefer. So the point is: Take charge by carrying your supplies with you.

Problem-Prevention Strategy #4

If you're vacationing somewhere and get caught by surprise, you might ask your mom to go buy supplies for you. That way, you can avoid being embarrassed by having to buy pads from the cutie who happens to work at the local store.

Plan Ahead

"One of my friends suddenly got her period, and bright, red blood went right through her shorts. She got all emotional and started yelling really loud. The people at school started looking at us. My friend finally calmed down and called her mom. Luckily, her mom brought her a change of clothes right away.

The thing was, my friend told me later that she kind of knew her period was coming, but she had ignored it because it was too much trouble. But look how much trouble ignoring it caused her!" — Ali

Problem-Prevention Strategy #1

Don't be tempted to ignore the fact that your period is due. Sometimes it seems like your schedule is beyond busy with schoolwork, sports, hobbies, and going places with your friends. The last thing you want to have to think about at times is your period. But there's no getting around it: By pushing it to the back of your mind and heading out the door without supplies, you're asking for menstrual mortification.

Problem-Prevention Strategy #2

Avoid going on emotional overload if a mortifying moment happens. By losing your cool and playing up the drama, you'll make the problem seem much, much bigger. You'll draw unnecessary attention to yourself and cause people around you to go on gossip overload. When you look back on the situation, you'll probably feel worse than if you'd just tried to handle things quickly and quietly.

Midday Disaster

"Once last year when I was in seventh grade, my period came in the middle of the day. It just happened to be incredibly heavy and within an hour and a half, my pad was soaked. I hadn't really ever experienced that heavy of a period before, and I didn't have another pad. I soaked through my pants. Although luckily it was near the end of the day and I could just dash to my mom's car so no one would see, I ended up leaving a big stain on the car seat. I was really upset. My mom didn't get all mad, though." —Emma

Problem-Prevention Strategy #1

As you already know, when you first start menstruating, your periods can be anything but regular. As a result, you need to be prepared for anything. That includes heavier flow moments. When you're bringing supplies, pack a few extra pads or tampons in your backpack. Better too many than too few.

Problem-Prevention Strategy #2

Stash an extra sweatshirt in your locker for emergencies. Not only will you be covered when you unexpectedly encounter cold weather, you'll also be covered in case of a menstrual emergency.

Problem-Prevention Strategy #3

Leaving behind a bright red stain on a chair or a car seat can definitely leave the most capable, cool girl feeling, well, *red-faced*! If your clothes are soaked and you're stuck having to sit somewhere, place your backpack or sweater under you. If worst comes to worst and you end up leaving a little blood behind, try to handle this calmly. The spot generally can be removed with cold water, special cleaning enzymes, or hydrogen peroxide. It isn't fun to deal with scenes like these, but it's definitely not the end of the world, either. Keep that in mind!

Final Words

By now, you've learned enough about your body to know that growing up is a wonderful, exciting, scary, complicated time of your life. You know that sometimes things will go along smoothly—and sometimes not-so-smoothly. But you're calm, cool, and collected, and you'll be prepared for changes when they come.

But you also know that it won't always be easy to "go with the flow." There will be times that'll test your girl power. There will be moments when you wonder just why you had to grow up at all. These moments might make you feel like you've lost control of your body.

At times like these, remember that you can take charge of your body by continuing to learn all you can about it. Your learning doesn't have to stop with this book or with health class. After all, it's your body. You're the one who reaps the benefits of a healthy self—and the one who has to deal with the consequences of bad decisions.

If you learn one important lesson from this book, it's to pump up your girl power by being an active participant in your personal health care. Always seek your doctor's advice and don't waste a moment seeking professional help if you have a concern. But remember that growing up means you can be a real teammate in your health care. YOU can ask questions. YOU can confide in your mom about your worries. YOU can tell your doctor everything about your concerns—without being embarrassed.

Congratulations! You're on your way to being a real woman—in every sense of the word!

Girl Stuff Glossary

ere is a short glossary of some of the most important terms you've read about throughout this book. By understanding the following words, you'll be able to explain some of the things your body is going through. You'll find that it'll be easier to understand body talk if you already know what these words mean.

endocrine glands: these glands release hormones into your blood and lymph system

estrogen: a female hormone that causes the eggs in the ovaries to "ripen" or mature

fallopian tubes: long, thin ducts that transport eggs (ova) from the ovaries to the uterus

genitals: organs pertaining to reproduction

gynecologist: a doctor who specializes in the health maintenance and diseases of women, especially the reproductive organs

hormones: chemicals that send messages through your body and are responsible for bringing about changes in your reproductive system

hypothalamus: the part of your brain that is a regulator and ultimately controls the menstrual cycle

menarche: the onset of menstruation (your first period)

menopause: the time in a woman's life when she stops menstruating

menstrual cycle: the process of ovulation, buildup of the lining in the uterus, and the monthly shedding of uterine material

menstruation: the discharge of blood and tissue from the uterus through the vagina

ova (singular: ovum): female reproductive cells

ovaries: the glands in which the ova develop

ovulation: the process in which eggs are released each month

period: the time of the month in which menstrual blood flows

pituitary gland: the gland affecting all hormonal functions of the body

premenstrual: the days leading up to the first day of your period, sometimes characterized by symptoms that the period is about to start

progesterone: one of the female hormones involved in menstruation

puberty: the time in life when your sexual organs mature

reproductive organs: the organs in your body, including the uterus, fallopian tubes, and ovaries, that are responsible for reproduction (creating babies)

secondary sex characteristics: changes including breast development and growth of pubic hair

uterus: the muscular organ in which a baby can grow

vagina: the opening through which menstrual blood and tissue is discharged

zygote: a fertilized egg (ovum)

Index